# Sustainable Community Movement Organizations

This volume shines a light on Sustainable Community Movement Organizations (SCMOs), an emergent wave of non-hierarchical, community-based socio-economic movements, with alternative forms of consumption and production very much at their core.

Extending beyond traditional ideas of cooperatives and mutualities, the essays in this collection explore new geographies of solidarity practices ranging from forms of horizontal democracy to interurban and transnational networks. The authors uniquely frame these movements within the Deleuzian concept of the 'rhizome', as a meshwork of alternative spaces, paths and trajectories. This connectivity is illustrated in case studies from around the world, ranging from protest movements in response to austerity measures in Southern Europe, to the Buen Vivir movement in the Andes, and Rotating Savings and Credit Associations (ROSCAs) in the Caribbean and Canada. Positioning these cases in relation to current theoretical debates on Social Solidarity Economy, the authors specifically address the question of the persistence and the durability of the organizing practices in community economies.

This book will be a valuable tool for academics and students of sustainable consumption, environmental policy, social policy, environmental economics, environmental management and sustainability studies more broadly.

**Francesca Forno**, Department of Sociology and Social Research at University of Trento, Italy.

**Richard R. Weiner**, Department of Political Science at Rhode Island College, USA / Minda de Gunzburg Center for European Studies at Harvard University, USA.

## Routledge Focus on Environment and Sustainability

**Design for Sustainability**
A Multi-level Framework from Products to Socio-technical Systems
*Fabrizio Ceschin and İdil Gaziulusoy*

**Sustainability, Conservation and Creativity**
Ethnographic Learning from Small-scale Practices
*Pamela J. Stewart and Andrew J. Strathern*

**Jainism and Environmental Politics**
*Aidan Rankin*

**Australian Climate Policy and Diplomacy**
Government-Industry Discourses
*Ben L. Parr*

**Reframing Energy Access**
Insights from The Gambia
*Anne Schiffer*

**Climate and Energy Politics in Poland**
Debating Carbon Dioxide and Shale Gas
*Aleksandra Lis*

**Sustainable Community Movement Organizations**
Solidarity Economies and Rhizomatic Practices
*Edited by Francesca Forno and Richard R. Weiner*

For more information about this series, please visit: https://www.routledge.com/Routledge-Focus-on-Environment-and-Sustainability/book-series/RFES

# Sustainable Community Movement Organizations

## Solidarity Economies and Rhizomatic Practices

**Edited by**
**Francesca Forno and**
**Richard R. Weiner**

First published 2020
by Routledge
2 Park Square, Milton Park, Abingdon, Oxon OX14 4RN

and by Routledge
52 Vanderbilt Avenue, New York, NY 10017

*Routledge is an imprint of the Taylor & Francis Group, an informa business*



*British Library Cataloguing-in-Publication Data*
A catalogue record for this book is available from the British Library

*Library of Congress Cataloging-in-Publication Data*
Names: Forno, Francesca, editor. | Weiner, Richard R., editor.
Title: Sustainable community movement organizations: solidarity economies and rhizomatic practices / edited by Francesca Forno and Richard R. Weiner.
Description: 1 Edition. | New York: Routledge, 2020. | Series: Routledge focus on environment and sustainability | Includes bibliographical references and index.
Identifiers: LCCN 2019058937 (print) | LCCN 2019058938 (ebook) | ISBN 9780367342234 (hardback) | ISBN 9780429324550 (ebook)
Subjects: LCSH: Community development—Environmental aspects. | Sustainable development—Moral and ethical aspects. | Sustainable development—Citizen participation. | Cooperative societies—Case studies. | Social movements.
Classification: LCC HN49.C6 S867 2020 (print) | LCC HN49.C6 (ebook) | DDC 338.9/27—dc23
LC record available at https://lccn.loc.gov/2019058937
LC ebook record available at https://lccn.loc.gov/2019058938

ISBN: 978-0-367-34223-4 (hbk)
ISBN: 978-0-429-32455-0 (ebk)

Typeset in Times New Roman
by codeMantra

# Contents

# Figures

# Tables

# Contributors

**Bengi Akbulut**, Department of Geography, Planning and Environment at Concordia University, Canada.

**Viviana Asara**, Institute for Multi-Level Governance and Development at Vienna University of Economics and Business, Austria.

**Alice Dal Gobbo**, Department of Sociology and Social Research at University of Trento, Italy.

**Francesca Forno**, Department of Sociology and Social Research at University of Trento, Italy.

**Paolo R. Graziano**, Department of Political Science at University of Padua, Italy.

**Caroline Shenaz Hossein**, Department of Social Science at York University, Toronto, Canada.

**Iván López**, Department of Psychology and Sociology at University of Zaragoza, Spain.

**Raquel Neyra**, Envjustice Project at Institute of Environmental Science and Technology at the Autonomous University of Barcelona, Spain.

**Angelos Varvarousis**, Institute of Environmental Science and Technology at the Autonomous University of Barcelona, Spain.

**Richard R. Weiner**, Department of Political Science at Rhode Island College, USA / Minda de Gunzburg Center for European Studies at Harvard University, USA.

# Acknowledgements

This anthology is dedicated to the commitment to developing new practices, to building new forms of social pacts from the bottom up. These are pacts developed in islands of alternative social solidarity economy in a capitalist sea as *sustainable community movement organizations* (SCMOs). These are forms of life which have blossomed rhizomatically in meshwork space, stressing horizontal power committed to intersubjective *rapports* in our reciprocal assistance. Geographies of co-responsibility emerge in managing both our mutual vulnerabilities to the vagaries of neoliberal dispositions and our creative capacity to build new institutional safeguards. Specifically, we are inspired by the embodied collective innovation in sustaining open-ended listening, deliberation and democratic experimentation in practice.

Both of us acknowledge the forums provided us, respectively, by the Society for the Advancement of Socio-Economics (SASE), the Council of European Studies, the European Sociological Association and the International Political Science Association.

Francesca is indebted to countless discussions on political consumerism and grassroots mobilization with Cristina Grasseni, Silvana Signori and all CoresLab researchers she worked with during her several years of teaching and researching spent at the University of Bergamo, where she also had the opportunity to follow the rise and development of "Cittadinanza Sostenibile" (Sustainable Citizenship), the local Solidarity Economy Network. Another forum I feel very indebted to is the SASE Network "Alternatives to Capitalism," established and coordinated by Torsten Geelan and Lara Monticelli. Over the years, this network has offered us a space for encounter and discussion to develop international, comparative and interdisciplinary studies on emerging new grassroots economic practices and associated institutions, making it possible for us to observe how similar projects are emerging, with similar characteristics and dynamics, all around the world.

Rich remains beholden to a seminar led by Trent Schroyer at the dawn of the 1970s that introduced him to the critical political sociology of social movements and contradictions of the welfare state elaborated on by Alain Touraine and Claus Offe. These interests became a calling that Robert K. Merton developed as he nudged Rich into working at theory of the middle range, grounded in testable empirical and historical fact. "RKM" is forever looking over my shoulder as I peck away at the keyboard. Further, I am thankful for the friendship of Mark Motte and Joe McCarney, enhanced not only by affinities but, more significantly, by a common sensibility of immanent critique, anticipating social forms of a better world. This anthology draws on the continuing stimulation of the Minda de Gunzburg Center for European Studies at Harvard University.

Trento, Trentino-Alto Adige/Südtirol, Italy
Providence, Rhode Island, USA

Chapter 2 is a re-elaboration of an article originally published in 2014 in the *Journal of Consumer Culture*, 14 (2): 139–157.

Chapter 6 is a slightly adapted version of an article published in 2015 in the *Journal of Co-Operative Studies*, 48 (8): 7–18.

# Glossary

**AENs** Alternative Exchange Networks (AENs) are solidarity-based exchanges and cooperative structures founded on principles of social and ecological values, participation and cooperation. Some examples are community currencies; barter clubs; citizens' self-help groups; farmers' markets; decentralized networks of workers' cooperatives; time banks; community supported agriculture; and various local community initiatives around food, housing and healthcare.

**Assemblage** Assemblage involves a putting together – as in a re-assembling – of previously disconnected practices, establishing relations between them. It is associated with *agencement*, the activity as well as the arrangement of an intensive ensemble or network, one constituted by emergent effects of self-ordering forces of heterogeneous material that come to mesh together. It prefigures and configures dynamics of becomingness amidst the *rapports* of intersubjectivity within an ensemble, and not within fixed structural relations. This involves a breaking up and then a recombinative participating in further connectingness. Alongside this, there is a capacity to interweave and enmesh horizontal/heterarchical connectingness.

**Buen vivir** Buen vivir (living well) is the Spanish term for the Indigenous worldview of *sumac kawsay*, i.e., ecologically balanced community life. It refers to a culturally sensitive and community-centric way of doing things: about not the individual *per se* but the individual in both a social solidarity context and a respective unique environmental situation. It is based on the belief that true well-being ("the good life") is only possible as part of a community. As a worldview, Buen vivir is a sensibility opposed to a development-centred approach. Furthermore, in its holistic view of life, humans are understood not as owners, but as reciprocating stewards of the Earth and its resources. Nature is therefore considered not as "natural capital" but as a being without which life does not exist.

**Commoning** Commoning defines a specific way of use and production, distribution and circulation of resources through democratic and horizontal forms of governance which give rise to and sustain social systems (the "commons") in which resources are shared by a community of users/producers (the "commoners"). As a verb, it describes the social practices used by commoners over the course of managing shared resources. Commoning means to participate in unfolding projects for social change, which aim to have implications for politics, economics and the planet.

**Deleuzian thinking** An *assemblage*-based ontology, Deleuzian thinking taps into the findings of theorizing on complexity and emergence. Such theorizing shows that critical thresholds in some physical and biological systems can be said to "sense" differences in their environment that trigger self-organizing practices. Gilles Deleuze and his frequent co-author Felix Guattari studied emergence in terms of potentiality rather than actuality, specifically the potential to differ. This potential is understood as one that is actualized in lines of creation and connectingness as well as in relations among geometries and geographies of power. Deleuzian thinking is more interested in a continuously recombinant becomingness and continued connectingness than in some underlying beingness. This becomingness is understood in planes of immanence. Immanence is here understood as lines of creativity/novelty, as fugitive moments that traverse a framed field, stir it up, deconstruct it, possibly leading to a coalescing around some new arrangement, that is, some new *assemblage*.

**Geographies of co-responsibility** Such geographies denote a wider sense of social-economic relationality where we encounter the complexity of engaging others in pluralized ever-changing struggles. Derived from the theorizing of Doreen Massey, the concept addresses multiplicities of spatial relations amidst both geometries of power and efforts at embodied co-responsibility. The co-responsibility looked to is grounded in spatio-temporal relations of connectingness, rather than in economistic definitions of maximization and optimization. Such a sense of geographies of responsibility involves fiduciary obligation as stakeholders within a trust network. This is a geographically imagined *assemblage* interweaving horizontal/heterarchical co-responsibility as a meshwork of value chains. Involved here is a pooling out of common goods, which were once called private goods.

**Micropolitical cartography** A sense of cartography that maps social movement of novelty. It is a *rhizomatic practices analysis* that we turn to in order to trace and portray alternative spaces, pathways, connections, conjunctions and trajectories "in between" assemblages and meshworks. As such, it taps into the capacity/capability involved in the mutual ordering of heterogeneous rhizomatic practices of connectingness. A rhizomatic practices analysis shifts focus from integrative functional differentiation in some organismic or systemic whole to performative connectingness in the emergent horizontality of innovative practice.

**Pooling resources** Pooling resources means inventing new forms of commoning based on the embodied inter-connectedness of reciprocal solidarity. This pooling is founded on social learning *vis à vis* local norms, values and interests at multiple scales. Through commoning/commons projects, bridges are built between and beyond social roles: for example, bridges between consumers and producers, employees and employers, and clients and service providers. Pooling resources of various types is fundamental to building new social systems in which reproduction stems from the direct participation of a community of users/producers who are able to reclaim and share. They are motivated by a set of values fundamentally opposed to those embedded in the capital circuits. These are values of solidarity, mutual aid, cooperation, respect for human beings and the environment, horizontalism and direct democracy.

**Rhizome** In Deleuzian thinking, rhizome is not an aborescent (tree-like) representation but a stringy/clewy-like substance. It is more of an immanent process of becomingness, connoting decentred multiplicity or network. Deleuzian thinking helps us grasp the unceasingly restless hyper-connectivity of cross-cutting, zigzagging and boundary spanning of performative offshoots. Rhizomes spin off their seeds like crabgrass: i.e., ceaselessly establishing connections, junctions, interconnections. These offshoots easily become woven, undone and re-woven. Rhizomatic reference denotes the subterranean character of the biological rhizome, evoking a network quality of interlinked forces beneath the surface that have adapted to resist striating forces from the surface and the air. As such, it is the opposite of Aristotelian concepts of rootedness/aborescence. Rhizomatic refers to the practice of rhizomes, not as foundational elements, but as ontological openness to continued connectingness, i.e., performative offshoots of cascading emergent connections and reconnections burrowing ever forward, capable of new elective affinities and upsurging.

**Rondas** The term relates to the phenomena in Peru of *Rondas campesinas*. These are community organizations of peasants in the foothills of the Andes. They are involved in alternative policing, monitoring and adjudicating of alternatives to practices associated with capitalist justice. The term Rondas comes from the Spanish for the practice of rounds of patrolling to keep watch against intrusions on communal territory, including its cattle and mineral resources. Completely horizontalist/mutualist in method, these practices became officially recognized and institutionalized in the Peruvian Constitution.

**ROSCAs** Rotating Credit and Savings Associations (ROSCAs) are informal financial institutions made up of a group of people which agree to put their money into a common fund, generally structured around monthly contributions. A ROSCA emerges within a community-organized series of contributions and withdrawals. ROSCAs are most (although not solely) common in areas where access to formal financial institutions is limited and among individuals who might not have access to such institutions.

**SCMOs** Sustainable Community Movement Organizations (SCMOs) are constituted by social movement actors who work towards building dynamics of innovation and reciprocity within communities. Within the SCMO, environmental protection and social justice issues are inextricably intertwined. Different than traditional social movement organizations, SCMOs have the peculiarity of mobilizing citizens primarily through their purchasing power. However, within these initiatives, the act of buying is promoted not simply individually but within a socialized sensibility among a group of people. This active and participatory collective practice qualitatively distinguishes its unique political action from individualized forms of political consumerism; SCMOs, in fact, utilize political consumerism not just to build awareness to step up pressure on producers and corporations but, even more significantly, to facilitate the construction of new alliances between different actors, starting from the local level.

**SEDs** Social Economy Districts (SEDs) are economic systems of material flows based on mutual engagement and joint activities among different actors that help each other to meet their needs of purchase, sale, exchange goods, services and information, according to principles inspired by a commitment to building an economy that is local, fair, supportive and sustainable.

**SSE** Social Solidarity Economy is an ethical and values-based approach to economic development. It is an approach that prioritizes the welfare of people and the planet, rather than profits and economic growth (as measured by gross domestic product [GDP]).

The core ideas of SSE are cooperation rather than competition and meaningfulness for people instead of profit. In such an approach, people play an active role in shaping all of the aspects of human life: economic, social, cultural, political and environmental. SSE as an approach can encompass all sectors of the economy – production, finance, distribution, exchange, consumption and governance.

# 1 Introduction

## Sustainable solidarity economies: rhizomatic practices for another world

*Richard R. Weiner and Francesca Forno*

### Introduction

Throughout the last decades, and particularly since the end of the Cold War and the spread of neoliberal capitalism, the global political space has profoundly changed. Shared disillusionment with traditional institutional politics has implied a retreat from its codified spaces; struggles for emancipation take place mostly through the informal constitution of groups that challenge traditional categories of political participation (e.g., left and right, class identities). Moreover, and not surprisingly, giving the market's increasing importance in shaping the (everyday) worlds of people across the globe, these movements move "from the streets to the market" – more and more often enacting politics through consumption (e.g., boycott, buycott platforms and apps, alternative/sustainable lifestyles). Furthermore, against a patterning of action that tended to verticalization and centralization in bureaucratic institutions, emphasis is given on decentralization, self-organization in non-hierarchical groupings and the creation of horizontal alliances of potentially global reach among local groups who further similar interests: they have a "glocal" dimension.

This anthology weaves together a coherent series of contributions and case studies on emergent social-economic forms of alternative organizing. These forms try to constitute autonomous normative ordering based on mutual regulating social-economic networks whose constitutive provenance lies in heterarchical multi-stakeholder social pact-ing. As such, they are both embedded social insertion and embodied responsibility of pooling common resources. These forms by following along the theoretical lines detailed by Elinor Ostrom (1990), may also be interpreted as inter-connectedness of reciprocal solidarity and endogenous trust for common resource stewardship.

Often also referred to as "common-based peer production" (or CBPP), such a form of collective action is emergent in different parts of Europe as well as in North and South America. By focussing not only on production but also in consumption, procurement, micro-financing renewable energy, organic food schemes, and anti-extractivism, they aim at evolving toward sustainable alternatives to commodified patterns of consumerism. In doing so they manifest intent on meeting the need of the present without compromising the capability of future generations to meet their own needs.

Within recent debate on collective action and various forms of activism, the concept of *Sustainable Community Movement Organizations* (hereafter SCMOs) has been proposed to indicate grassroots efforts to build alternative, productive, and sustainable networks of production and exchange by mobilizing citizens primarily (though not solely) via their purchasing power (Forno and Graziano, 2014). By acting primarily on the market, such grassroots initiatives attempt to create new economic and cultural spaces for civic learning as well as consumerism and producerism actions (Andretta and Guidi, 2017) that aim to construct and sustain alternative markets based on knowledge exchange, loyalty, and trust. In other words, the networks they form facilitate both the circulation of resources (information, tasks, money, and goods) and the construction of common interpretations of reality, thus simultaneously providing a framework for collective action and enabling the actual deployment of alternative lifestyles (Forno *et al.*, 2015).

SCMOs often include experiences of mutualism, such as in projects of welfare from below, consumer-producer networks and cooperatives, Alternative Food Networks, urban agriculture/urban gardening, barter groups, time banks, recovered factories, local savings groups/alternative currencies, fair trade, ecovillages, and social and solidarity economy networks. These initiatives address at once *ecology* (climate change, resource depletion, reduced biodiversity, diminishing land fertility, diminishing wildlife, etc.); *economy* (reduced family's income and purchase power, unemployment, increasing difficulties for small enterprises to keep afloat in the face of increasingly powerful and oligopolistic multinationals); *society-culture* (insecurity and unsafety, polarization of life opportunities, diminishing happiness and well-being, spreading of so-called psychological discomfort) in their being interconnected and co-emerging aspects of the same system.

Instead of appealing to formal (local, national, international) institutions by lobbying and/or putting pressure so as to make them change their political decisions, SCMOs act locally by ongoingly building concrete alternatives to the system they are contesting. Instead of asking for change they produce the change itself in the form of alternative ways of socio-ecological and economic organization, establishing novel material and cultural-symbolic patterns.

The cases discussed in this anthology demonstrate how SCMOs are alternative organizations which, while contesting around capitalism and markets, experiment with alternative ways of organizing. Then is done in the attempt to revamp moral principles (such as equality, democracy, and sustainability) within society and to contrast growing extremism and populism sentiments. By foreshadowing a confederal frame of thinking, imaginatively projecting and anticipating a more flexible and polycentric institutional architecture, such efforts involve a re-territorialization and re-municipalization of material flow in interlocal/interurban networks which more and more often converge in transnational advocacy networks of cities pursuing a geopolitics of communing and code connectivity.

SCMOs are the bedrock within the so-called *Social Solidarity Economy* (SSE), which is a term increasingly being used to refer to a broad range of organizations that are distinguished from conventional for-profit enterprise, entrepreneurship, and informal economy as they have explicit economic, social, and environmental objectives (Utting, 2014). All those myriad of experiences include cooperatives, mutual associations, NGOs engaged in income generating activities, women's self-help groups, community forestry and other organizations, associations of informal sector workers, social enterprise, and fair trade organizations and networks. Such movement organizations are understood as the basis of:

- a *sustainable governance*, and
- generative social entrepreneurship of stakeholders co-creating value by a confederated communing connected to networks.

As Forno and Graziano note in the second chapter of this anthology, SCMO self-reinforcing value chain networks inclusively interlace and encourage direct relationship between consumer and producers. For example, there are innovative micro-credit/micro-financing alternatives; local and social currency alternatives; and investments through purchase of shares/or deposits with ethical financed institutions linked to another new practice, "ethical banking." SCMO institutionalized governance is organized, not by private capital, but by mutualizing and networking cooperatives aligned SSE commitments and practices based on Alternative Exchange Networks (AENs) and/or Solidarity

Economy Districts (SEDs). In such governance, reciproqueteurs associate themselves on the basis of skill and through effective communicative interaction on a multiscalarity of platform levels.

SCMOs are motivated by commitment to the creation of shareable resources along with the democratic governance of such resources to sustain people and the planet. They use markets and their privileged limits – rather than the streets – as the battlefields. But beyond the governmentality of neoclassical and neoliberal market-mindedness, SCMOs move to the creation of what Ostrom refers to as *non-rivalrous common pool resources*, both alongside and outside the market. These constitute eco-systems evolving toward sustainable development and alternative forms of consumption in different parts of the world.

Concretely, the SCMO involves a new mode of resilient social pact-ing that is grounded in *decentred mutual stakeholder social pacts* (MSSPs). The intention of such new multi-stakeholder social pacts is to rebuild new social relationships around some radical revision of the market and practices of an excessively commodified environment (Bäckstrand, 2006). SCMOs are institutionalized by MSSPs locally, and beyond in levels of multi-scalarity.

The SCMOs' purpose is the creation of *value chains* of shareable resources embedded within the system of capitalism, such as knowledge, services, goods, labour, and solidarity purchasing power. Within this value chain network are co-stakeholders, co-producers, co-owners, co-users, and co-responsible *reciproqueteurs*. They seek to improve market access and market opportunities, therefore addressing grand challenges such as tackling climate change, fostering gender equality and attempt to reduce poverty, providing good food for all and more affordable healthcare.

In other words, the SCMOs can be understood as a *community of practices* underpinning new forms of consumption and production: an enduring constellation of arranged interconnected performances, with an arc of subject positioning (Torfing, 1999), a trajectory of how such practices shape, spawn, and develop each other. The SCMOs are *assemblages* of interwoven practices of mutually interactive performances for the sake of commitments, expressing mutual accountability (see here Rouse, 2006: 333; Schatzki 2008: 33). In this sense, they are "prefigurative": they try to embody an alternative world that might become concrete for people outside of their "niche." Privileging direct interaction and rejecting higher-level organization, in fact, does not imply that they cannot scale up through replication, the creation of networks, and alliances.

Stunning us out of habituated narrative, these (pre)fugitive instances capture moments of rupture and a collective imaginary, which overflow the framed field and mobilize demands that cannot be sufficiently satisfied in a habituated present (Touraine, 1977: 362). There is an institutionalizing at work – as an imaginative projecting of new values (Joas, 1993, 2000a, 2000b). This projecting opens up a constellation – an arc, so to speak – of intersubjectively new horizontal social positioning. And with it, a future of difference and heterarchy, rather than of homogenizing hierarchical ordering (Torfing, 1999). New spacings and new timings are opening in our midst in geographically dispersed meshworks (Delanda, 1996, 1998, 2016) across conventional territorial and organizational boundaries.

## SCMOs as rhizomatic becoming

SCMOs emerge and evolve as the result of long-term social practices by creating new "spaces" or "fields." In doing so resound of echoes with Gilles Deleuze and Felix Guattari's (1987) ontològy of creativity and becomingness. This is the swerve of novelty, prescient fugitive flashes of fugitive moments sublimating into institutionalizing form (Wolin, 1994) which – the subtitle of the anthology refers to. In order to map the complexity of immanent relations that are not reified and fixed into something neatly defined, Deleuze and Guattari introduced the concept of *rhizome*. As a concept, rhizome helps to grasp the unceasingly restless hyper-connectivity. This is the cross-cutting, zig-zagging boundary-spanning of performative offshoots. This is the incessant connecting, which includes all concerned actors involved in the on-going process of making participation and negotiation[1] (see Deleuze, 1986).

As opposed to Aristotelian concepts of rootedness and aborescence, Deleuze and Guattari approach sensitizes us to comprehend the cross-cutting, zigzagging, entangling undergrowth burrowing ever forward. What is described is path-creative action of emergent connecting-ness and weaving/enmeshing. Deleuzian thinking is more interested in their becoming open to continued connecting-ness than in their being (van Wezemael, 2008: 170–171).

### *Root gives way to rhizome*

What endures, noted Carl Jung (1965), "endures beneath the eternal flux. The plant living in its rhizome. Life hidden in *the rhizome*." Within the available spaces in fissures, rhizomes spin off their seeds like crabgrass, ceaselessly establishing connexions, junctions, interconnexions – woven, undone, rewoven. Rhizomes wander imaginatively, mangling and entangling. And we can detect map-like folding and refolding – open, reversible, detachable, reworkable in constant iteration and alteration.

The perspective from rhizome therefore enables to appreciate the complex dynamic of SCMOs emergence, as well as the multiple nodes of unceasing creation and cross-connexion they establish while building new heterarchical multi-stakeholder social pact-ing. The rhizome is a practice/ a flow, not a structure. It calls forth an entangled web of movement as displacement and transition. Specifically, not understanding society as a structure, but as a clustering of self-spreading flows, capable of new elective affinities. *The rhizomatic* connotes upsurging, wandering, and dispersal – emergent properties of connecting-ness, rather than fixed ones. Rhizomatic interlinkages can be disruptive fostering of novelty or a recursive fostering of stabilization.

Rhizomatic emergence relates to a clustering and interweaving, a capacity to assemble and enmesh heterarchical entangling rather than hierarchical embedding. This is understood as horizontally oriented co-articulation, co-responsibility. What Deleuzian thinking has come to refer to as *assemblage.* De Landa (2016: 1) defines this term as a multiplicity which is made up of many heterogeneous terms and which establishes liaisons, relations between them. In so doing, actors and actants make emergent sense of each other, and not simply by the summation of the properties of their components.[2] An assemblage of practices and discourse configure and prefigure a becomingness within dynamics amidst *the rapports* of the ensemble, not within fixed structured relations (again, see Touraine, 1977). There is a re-assembling of previously disconnected practices.

Assemblage involves breaking up and then recombinatory participation in further connecting-ness and a capacity to enmesh that connectingness. As social insertion, there is an opening up of closed circuits of supposedly stable regulation before supposed structures unravel and dikes collapse. Rhizomatic assemblage amounts to an enmeshing composite of elective affinities, articulated as a rather stable operationally autonomous ensemble of interdependent and mutually reinforcing practices, whose participants interact through negotiations. In Deleuzian writing, assemblage is associated with *agencement*, the activity as well as the arrangement of the ensemble itself (Phillips, 2006).

Such an ensemble is referred to as a *meshwork* of connecting-ness that recursively interweaves and coordinates heterogeneous actants in a self-sustaining connecting-ness. Such meshwork has a fibrous, clewy, stringy-like character. These are force-fields within the nodes that materialize with as many dimensions as they have connexions (Latour, 2007). The force-fields exhibit an enmeshed pattern of negotiations while at the same time manifesting intensive processes of discontinuous knowing, stirring, and emergence. Thus enmeshed, assemblages manifest emergent effects. Rapports amount to felt immersion in connecting-ness, an active sense of living by intersubjectively acknowledging one another as co-authors of idea, projects, and institutionalizing practices. This is an intersubjectivity of our creative interactions as a mutually lived experience – as a shared decoupling, recoupling, and co-emergence. Rhizomatically, rapports are scrambled, no longer conforming to the subject position spacings of the past or the present. This is a *rhizomatic emergence* facilitating norms of connecting-ness, a regulative normative framework that is to a certain extent self-regulatory.

In other worlds, there is in Deleuzian thinking a sense of continuously recombinant twists and turns, a sense of experimenting with practices, and creating new values. This is a sense of "lived planes of immanence" within practices, but it is not an ontology of internal relations with either teleology or transcendence. These are planes of becoming-ness, planes of actualization. Agency is immanent within a forming assemblage rather than in some subsuming totality, no longer understood as some universalizing grand narrative (Phillips, 2006). Deleuzian thought engages us to consider becomingness without falling into a linear realization of the possible. What is described is cascading path-creative action of emergent connecting-ness and re-connecting-ness, rather than seeming determined, sequenced synchronic path-dependency.

As "vital forces" which are virtual and actual at the same time, SCMOs can be comprehended rhizomatically, as an assemblage of constituting practice and discourse. Processes of "becoming" that can be studied as intertwined movements generative of a new way of being, ones that prefigure and configure a new sense of bindingness. Accordingly, sociological discovery needs to concentrate on understanding to what extent does such an enmeshed ensemble operate with some consistency and coherence, trying to map how such rapports scrambled in a zigzag endless movement and how then, such movement involve decoupling, recoupling, and track-switching.

Stringy/clewy rhizomatic interconnexions of solidarity practices comprise innovative valuation, innovative value chains as well as supply chains in solidarity economies. The interconnections are immanent substance of material practices resisting and disembedding conditions of commodification. All patterns that are understandable as the imaginative and institutionalizing projection of new values reveal a predicate logic of intersubjective practical reasoning. Further, as Brandom (1979, 1983, 1995) describes, what they reveal are the normative pragmatics involved. And they are understood as chains of unfolding signifiers, unfolding categorically within a context of situated agency.

As the case studies discussed in this volume demonstrate, SCMOs can be understood as rhizomatic emergence, as well as a trajectory of a multiplicity of connexions without fixed subject positioning, and with interwoven heterogeneous paths of substantive norm-setting. Their creative becoming-ness can be comprehended as immanence toward sustainable development, a new sense of bindingness – a sense of reciprocal solidarity regarding sustainable development marked by immanent claims, what Robert Brandom (1979, 1995) refers to as *assertional warrants* with normative commitments, bringing something new into the web of entangled values and norms as moral practice. From this point of view, organizations – as socially constructed entities – form or modify through an on-going process of making participation and negotiation, which happens within the social context in which they are embedded in and through a complex relation of knowledge and power.

## Analysing SCMOs with a rhizomatic lens

From the enigmatic philosophical insight of Deleuze and Guattari it is possible to glean an analysis of such performative social practices that we refer here as Rhizomatic Practices Analysis (RPA). RPA indicate the need to shift our focus from integrative functional differentiation in some organic whole – to the performative *connectingness* in some emergent horizontality of innovative practices. Rhizomes are not arborescent/tree-like rooted foundational elements, but performative offshoots. Following Deleuze and Guattari in fact, society is not to be understood as sustained by constructed pillars of corporatism, but as shifting assemblages without a real centre (see Deleuze and Parnet, 1987). As rhizomatic systems evolve through process of problematization, becomingness needs to be approached as a process-mapping without falling into a process-tracing of linear realization of the possible. Less in terms of network innovation, and more along the lines of Bourdieu's study of a field (Bourdieu, 2000). Such a field is not just one of disruptions, transgressions, and collisions but also, at the same time, one of patterned flows of traceable unfolding institutionalizing practices themselves.[3]

RPA therefore aims at mapping ontological emergence in assemblage/meshwork, and clustering relays of assemblages horizontally connecting multiple points amidst paranodality. As understood by Guattari and Deleuze, the rhizome itself is a map of lines of disruption transforming into lines of flight and re-connecting-ness; its maps exhibit a process of relational and transversal meshworking. RPA mapping suggests, therefore, that we focus on the constantly contesting character of our social existence – contested structuring processes, rather than on structures themselves. This is more than process-tracing of multiple data points in political science input-output analysis. The very idea of representation is de-stabilized.

Sociological investigation into SCMOs should consequently move beyond bounded rationality to understand restless hyper-connectivity, how our innovation processes will always outrun our social representations. As a sort of micropolitical cartography of alternative spaces. RPA suggests that we follow the paths and trajectories "in between" assemblages and networks. Thus, RPA aims to tap into the capability revealed in mutual constitutive ordering. Specifically here, the capability revealed in heterogeneous rhizomatic practices of connectingness.

RPA journeys into the interior of bearing signification, into our constituting values, norms, and commitments. Practices are predicated on a culture of deliberative horizontality, exhibiting a social bondingness and commoning practices.[4] Rhizomatic emergence should in fact not just be understood as disruptive and transgressive, but also as creative and constructive: opening patterns of self-differentiation and self-ordering beyond a micropolitics of affinity. Stabilization should therefore be described as unhinged, fractured, de-coupled and re-coupled within lines of immanence, within trajectories of new valuation.

Comprehensible as argumentative exchanges with warrants that swell and overflow frames of reference as path-disrupting and path-creating collective action, RPA sensitizes us to ontological emergence – to comprehending the perturbations, eruptions, collisions, contestation, committedness, and surges. RPA therefore helps to gauge the learning process involved in how these surges *in formation* generate new "forms of life" – opening up new circuits of re-regulation, new patterns of self-ordering as reflexive governance (Voss *et al.*, 2006).

More deeply, Rhizomatic Practices Analysis – like ethnomethodology and phenomenological sociology (e.g., Berger and Luckmann, 1966; Garfinkel, 1967; Schutz, 1972) – should aim to produce thick description of practices with normative commitment. This is a thick description of their participating in promising practices with associated accounting practices and interpretive procedures. However, while the RPA proposed here shares intertwining affinities with the constructivist indexical expression analysis of ethnomethodology and phenomenological sociology, it nevertheless avoids their common perspective of "bracketing," which Alfred Schutz endows to them – the bracketing of issues of power generating the "constantly contesting" lifeworld. Knowledge should in fact to be understood and traced as continually *in-the-making*, that is, in a continual negotiation, renegotiation and reconstruction of shared conceptualizations (e.g., Moscovici, 2001).

RPA should therefore be recognized as a mode of investigation that is rooted in time and place. A way to conduct research that needs a multiplicity of methods, using analysis of historical records and documents as well as formal and informal conversations. As focussed on becomingness within the unfolding of innovation as on the interconnectingness of paranodality, RPA implies the need to divert from usual analyses based on linear or binary associations to concentrate on "lines of flight," which, in Deleuze's words, traverse a framed field, stir it up, deconstruct it, and prevent its closure (1986).

RPA aims to grasp how rhizomatic patterns evolve within an autonomous social-economic form that amounts to a meshwork of mutual responsibility and social insertion. Its ultimate concern should therefore be to discern: (1) how mutually interactive performances are accountable to each other; (2) how connecting rhizomatic practices are brought together. Put differently, RPA should tap into the capability of mutually constituting orders of heterogeneous forces to mesh together. RPA needs to understand how disparate rhizomatic sub-meshworks are brought together with some recurring coherent coordination as an enduring pattern of participatory governance. What is asked: How are they brought together in a manner enabling them to mutually self-order and self-differentiate themselves as reflexivity, and then to spread interwoven meshworks of horizontal reciprocal solidarity? Novel valuation is understandable as meaning construction in terms of connectivity norms, coherency norms, capability norms.

Specifically, taking a RPA perspective means to:

- grasp moments of rupture;
- apprehend the transgressing of inherited normative forms;
- denote transversal movements of overlapping and cross-cutting nodes, junctures and inter-connexion, as well as their capability/capacity to assemble, interweave, and enmesh connexions "in between";
- capture movement as a flow swelling beyond, overflowing and displacing the frames which movement transgress and disrupt;
- discern how practices and discourses merge and interweave with each other to form a discursive web and make plausible and meaningful sense;

- eschew any drawing upon either reductionist sort of fixed constellation of unities, or snapshots of periodized moments of time;
- take hold of the unfolding of both the flow and the predicate logic of movement becoming institutionalizing practice;
- to map and trace "lines of flight"/"planes of immanence" opening up "planes of possibility" as fugitive frames of intersubjectively imagined new social positioning;
- interpret how these fugitive frames work at prefiguring institutionalizing practice and co-producing a form of participatory governance;
- discern how alternative valuation emerges and is offered up as an act of creative capability to participate in a movement of becoming.

## Outline of this collection

The series of works presented in this anthology will show how in a social-economic sense, SCMOs take us beyond 19th century and early 20th century forms of syndicalism, municipalism, and cooperatives. By addressing grand challenges for our societies as they are building alternative, productive, and sustainable networks of production and exchange, *SCMOs* urge us to develop novel avenues of critical thinking to comprehend how rapports among and between such forms of organizing are comprised.

Overall, the different cases presented here discuss:

i the emergence of new geographies of solidarity practices from municipal horizontal democracy to interurban and transnational networks, as well as how they are interlinked in different ways;
ii investigate how and in what ways SCMOs communicate and set up relations within AENs;
iii how such culture of horizontal networks and connectedness are gauged;
iv how both ideas and practices of co-responsibility and reciprocal solidarity embody and sustain emergent SCMOs;
iv test metaphors of SCMOs, AENs, and SEDs as distinguishable "islands of alternatives in a capitalist sea" specifically the extent to which they resist conditions of commodification and disembedded atomism;
v raise questions regarding the durability and resilience of these organizations' dedication to commoning and pooling resources as they move into power at the municipal level;

vi query to what extent rhizomatic assemblages of cross-cutting connectedness emerge when people resist processes of privatization and deregulation, claiming the factor as a commonality through affective politics of precarity, despair, and co-responsibility;
vii query to what extent SCMOs are focussed continuations of both the Global Justice Movement and the Degrowth Movement related especially to food, finance, extractivism, and climate change;
ix query to what extent do sustainable solidarity economies' practices take us beyond Elinor Ostrom's focus on the rules and institutions governing the commons;
x query to what extent can we create more linkages and networks between The North and The South when the multinational corporations of The North and China are such a part of the pattern in The South, feeding the cycles of "unsustainable consumption" in The North.

In the first case-based study, starting with our second chapter, Francesca Forno and Paolo Graziano draw on their foundational article on SCMOs published in 2014. They underline how, in the current time characterized by a "multiple crisis," social movements are simultaneously facing two types of challenges. First, they are confronting institutions which are less able (or willing) to mediate new demands for social justice and equity emerging from various sectors of society. Second, given the highly individualized structure of contemporary society, they are also having trouble in building bonds of solidarity and cooperation among people, bonds which are a fundamental resource for collective action. SCMOs are practice-based movement actors whose main aim is to bring different collectives together to help them developing a post-capitalistic socio-economic system in which the overriding object of profit maximization is substituted by cooperation, solidarity, and mutualism. While sharing several common traits with social movement organizations of the past, such actors tend to bypass the traditional state – addressing repertoires of action, and to focus on changing society as part of everyday politics, where the public and private spheres are increasingly blurred. Although with some differences due to their specific geographical origin, as highlighted in this chapter, SCMOs share several common traits regarding their motivations, their repertoire of action, and their organizational structures.

In the third chapter, Richard R. Weiner and Iván López take the discussion on SCMOs further developing an operationalization of such new social-economic concept of embodied economic co-responsibility, resource pooling, and stakeholder stewardship. As formulated, SCMOs illuminate a new form of social insertion: the creating from the bottom-up of sustainable value networks by co-producing stakeholders. What is projected and constituted discursively and economically is trust networking with which to frame and reconfigure institutional practices. SCMOs co-construct a sensibility of a need for a river going back to its normal flow after a disastrous flood, as in unbridled growth and capital accumulation. Such trust networking involves paranodality, the involvement of more than one single dominating code. The chapter utilizes a case-study focus on the lessons learned from the 15M movement in post-Franco Spain, when new initiatives emerged as new spaces for social action. The abbreviated name "15M" refers to the date of May 15, 2011, when thousands of people – mostly from the 50% plus unemployed young people – occupied public spaces in major Spanish cities, especially in the well-known Puerta del Sol in Madrid. Having lost trust in the two-party system, this movement targets embedded regime corruption and demands more citizen participation. Further, it has developed autonomous SCMOs for solidarity-based exchange on an urban and inter-urban basis.

In the fourth chapter, Angelos Varvarousis, Viviana Asara, and Bengi Akbulet explain how the "movement of the squares" has produced a vast literature, where most of the attention has focussed on the encampments period. The case-based study examines the movement's unfolding following the end of the more visible cycles of mobilization and its decentralization. It is argued that a crucial aspect of this evolution lies in the creation of a social infrastructure of alternative (re)productive projects. They call this type of outcomes "social" in order to distinguish them from the cultural, political, and biographical outcomes underlined in typologies on the consequences of social movements. Through a comparative analysis of the movements in Athens and Barcelona, they show how the commoning practices of the square encampments gave rise to more enduring commons disseminated across cities′ social fabrics. Their analysis identifies both direct and indirect mechanisms of movements′ transmutation into commons. Further, the authors distinguish the former into transplantation, ideation, and breeding processes. The article also scrutinizes the political dimension of these commons in relation to what has been framed as the "post-political condition." Ultimately, it is maintained that the post-square commons constitute political and politicizing actions for activists, as well as for users of the effects of commoning on everyday life. The authors discern the capacity of such commoning in linking their practices with broader, structural dynamics of injustice, inequality, and exclusion; as well as with their selective engagement with counter-austerity politics. This paper constitutes one of the scant empirically grounded attempts to bring together social movement studies with the literature on commons, and to build a conceptual framework of their relationship within a rhizomatic lens.

In the fifth chapter, Alice Dal Gobbo and Francesca Forno examine how overconsumption, with associated processes of consumerism and commodification, has gained centrality and it is today at the heart of several contemporary social movement organizations which stress the contradiction between capitalist growth versus living conditions in the community. Drawing on the case study of GAS (Solidarity Purchase Groups) in Italy, this chapter looks at how people self-organize to achieve socially and environmentally sustainable transitions. Key issues addressed in this chapter are how political consumerism and collective practices of sustainable procuring and provisioning can challenge commodification. Further, these SCMO practices constitute the bases for original "assemblages" that not only give concrete alternatives to the current unsustainable system, but indeed embody a novel style of doing politics.

In the sixth chapter, Caroline Shenaz Hossein details how millions of Black people in Jamaica, Guyana, Trinidad, Haiti, and Toronto use informal cooperative banking systems in low-income communities known as ROSCAs. These are money pools embedded, organized, and managed by women known as banker ladies on the basis of peer to peer (P2) lending alongside conventional commercialized banking systems. This is not done only to meet livelihood needs, but also to build reciprocity by helping their family, friends, and community. This chapter utilizes interviews and focus groups held in the aforementioned Caribbean and Canadian sites with 322 people between 2007 and 2015. Money pools function as sustainable solidarity economies evolving from ancient African traditions. They ripen under conditions of slavery and colonization. This case study research argues that throughout the Caribbean, and in the diaspora to Toronto, Indigenous ROSCA banking systems – with localized names such as susu, partner, meeting-turn, box-hand, and sol – are enduring rhizomatic practices that historically and currently are taking a bold stand against exclusionary financial systems.

In the sixth chapter, Raquel Neyra starts with how driven by economic growth policies, material and energy extractivism in the Global South offers cheap uprooting of natural resources and labour prices. Then Neyra goes on to detail how in the Andes, capital accumulation meets with local resistance of peasants and Indigenous peoples. Neoliberal policies of Peru since the 1990s emphasize extractivism as the foundation of all the future governments. In Peru, we have resistance of "Rondas Campesinas" and the Defence Fronts. When peasants and Indigenous peoples defend their habitat and way of life, they participate in the preservation of the "buen vivir" ("living well"). The defence of the environment gives them a new sense to their lives, a sense that they draw on a recovered culturally sensitive world-view and material practice known as "buen vivir."

The solidarity economy practices of SCMOs rhizomatically constitute not only geographies of co-responsibility, but constitute as well ecosystems evolving toward sustainable albeit alternative modes of consumption and production in different parts of the world. These practices move from urban to interurban meshworks, and then cross national borders through transnational advocacy networks (TANs). These case studies are rich and at times both unheralded and unexpected as they catalogue and delineate a new sense of bindingness and autonomy in these SCMOs' innovative connected collective actions. There is an underlying connectingness here, one that reveals how these SCMOs project new values and new practices – to open up closed circuits, overflow them, and lead us beyond the grip of persisting neoliberal governmentality.

## Notes

1 In *Cinema 1* (1986), Deleuze introduces the concept of "planes of immanence" as the medium through which we creatively evolve. These planes of immanence denote fugitive moments – "lines of flight" – that traverse a framed field, stir it up, deconstruct it, and prevent its closure. Each line of flight is a path taken to flee a given arrangement of some bundle of practices, possibly leading to that bundle's disillusion, possibly coalescing around some new arrangement. The "lived planes of immanence" within practices are not just lines of flight, but planes of becoming-ness.

2 DeLanda (2006) describes Deleuzian thinking as assemblage-based ontology, in terms of Leibnizian compossibility. That is ontology that is open to continued connecting-ness, so long as the context of innovation and novelty are sustained in a non-essentialist manner, and continuously re-made. Latour (2007), in *Reassembling* the Social, speaks of a clustering of self-spreading assemblages, self-spreading flows capable of new elective affinities.

3 Alexandra Steinberg discusses "rhizomic network analysis" more in terms of Bruno Latour's Actor-Network Theory (ANT) where networks as actants interact with one another than the more Bourdieusian RPA approach discussed here. Steinberg, "Rhizomic Network Analysis: Toward a Better Understanding of Knowledge Dynamics of Innovation in Business Networks," in Fang Zhao, ed, *Handbook of Research in Information, Entrepreneurship and Innovation* (Melbourne: Idea Group, Inc., 2007).

4 This is suggested by Elinor Ostrom's ideal typical reciproqueteurs pooling resources. For Ostrom, commoning involves founding and enforcing institutions for governing knowledge and resources. Ostrom discussed institution-building to govern the commons as "polyvalent" and multi-scalar – similar to assemblage/meshwork theorizing. However, her analysis was functionalist in character as it focusses on the roles and institutions "governing the commons" – not on the process of commoning itself, not on the process of commoning in terms of social solidarity. As such, it is closer to a rational institutionalist account than the RPA approach we are developing here.

## Bibliography

Andretta, Massimiliano and Guidi, Riccardo, 2017. "Political Consumerism and Producerism in Times of Crisis. A Social Movement Perspective?" *Partecipazione & Conflitto*, 10 (1): 246–274.

Bäckstrand, Karin, 2006. "Multi-Stakeholder Partnership for Sustainable Development: Rethinking Legitimacy, Accountability, and Effectiveness." *European Environment*, 16 (5): 290–306.

Berger, Peter and Luckmann, Thomas, 1966. *The Social Construction of Reality*. Garden City: Doubleday.

Bourdieu, Pierre, 1992. *The Logic of Practice*. Trans. Richard Nice. Stanford: Stanford University Press.

Bourdieu, Pierre, 2000. *Propos sur le champ politique*. Lyon: Presse Universitaire de Lyon.

Brandom, Robert, 1979. "Freedom and Constraint by Norms," *American Philosophical Quarterly*, 16 (3): 187–196.

Brandom, Robert, 1983. "Asserting." *Noûs*, 17 (4): 637–650.

Brandom, Robert, 1994. *Making It Explicit: Reasoning, Representation and Decisive Commitment*. Cambridge: Harvard University Press.

Castells, Manuel, 2016. *Networks of Outrage and Hope: Social Movements in the Internet Age*. Cambridge, UK: Polity Press.

De Landa, Manuel, 1996. *A New Philosophy of Society*. London: Bloomsbury.

De Landa, Manuel, 1998. "Deleuze and the Open-Ended Becoming of the World." Paper presented at Chaos/ Control: Complexity Conference, Bielefeld www.cdc.vt.edu/host/delanda/paper/brcoming.html

De Landa, Manuel, 2016. *Assemblage Theory*. Edinburgh: Edinburgh University Press.

Deleuze, Gilles, 1986. *Cinema 1, The Movement Image*. Trans. Hugh Tomlinson and Barbara Habberjam. Minneapolis: University of Minnesota Press.

Deleuze, Gilles, 1991. *Bergsonism*. Trans. Hugh Tomlinson. Cambridge: MIT Press.

Deleuze, Gilles, 1992. *The Fold: Leibniz and the Baroque*. Trans. Tom Conley. Minneapolis: University of Minnesota Press.

Deleuze, Gilles, 1997. *Negotiations, 1972–1990*. Trans. Martin Joughin. New York: Columbia University Press.

Deleuze, Gilles and Guattari, Felix, 1987. *A Thousand Plateaus*. Trans. Brian Massumi. Minneapolis: University of Minnesota Press.

Deleuze, Gilles and Parnet, Claire, 2007. *Dialogues II*. Trans. Hugh Tomlinson and Barbara Habberjam. New York: Columbia University Press.

Forno, Francesca and Graziano, Paolo R., 2014. "Sustainable Community Movement Organisations." *Journal of Consumer Culture*, 14 (2): 139–157.

Forno, Francesca, Cristina, Grsseni and Silvana, Signori, 2015. "Italy's Solidarity Purchase Groups as 'Citizenship Labs,'" in Emily Huddart Kennedy, Maurie J. Cohen, Naomi Krogman, eds. *Putting Sustainability into Practice: Advances and Applications of Social Practice Theories*, Northampton: Edward Elgar, pp. 67–88.

Garfinkel, Harold, 1967. *Studies in Ethnomethodology*. Englewood Cliffs: Prentice Hall.

Garud, Raghu, Kumaraswany, Aran and Karnøe, 2010. Peter, "Path Dependency or Path Creation," *Journal of Management Studies*, 47 (4): 760–774.

Habermas, Jürgen, 2001. *On the Pragmatics of Social Interaction*. Trans. Barbara Fultman. Cambridge: MIT Press.

Habermas, Jürgen, 2003. *Truth and Justification*. Trans. Barbara Fultman. Cambridge: MIT Press.

Harvey, David, 2012. *Rebel Cities: From the Right to the City to the Urban Revolution*. London: Verso.

Hasker, William, 1999. *The Emergent Self*. Ithaca: Cornell University.

Joas, Hans, 1993. *Pragmatism and Social Theory*. Chicago: University of Chicago Press.

Joas, Hans, 1997. *The Creativity of Action*. Chicago: University of Chicago Press.

Joas, Hans, 2000a. *The Genesis of Values*. Chicago: University of Chicago Press.

Joas, Hans, 2000b. *Praktischen Intersubjektivät*. Frankfurta. M.: Suhrkamp.

Jung, Carl, 1965. *Minds, Dreams and Reflections*., Ed. A. Jaffé / Trans. R. Winston and C. Winston., New York: Random House Vintage.

Kim, Jagwon, 1999. "Making Sense of Emergence," *Philosophical Studies*, 95 (1–2): 3–26

Latour, Bruno, 2007. *Reassembling the Social: An Introduction to Actor-Network Theory*. Oxford: Oxford University Press.

Massey, Doreen, 1995. *Spatial Divisions of Labor*. 2nd Ed., London: Routledge.

Massey, Doreen, 2004. "Geographies of Responsibility." *Geograpfiska Annaler*, 88: 5–18.

McCarthy, Joseph D. and Zald, Mayer, 1977. "Resource Mobilization and Social Movements: A Partial Theory." *American Journal of Sociology*, 66 (6): 1212–1241.

Moscovici, Serge, 2001. *Social Representations: Explorations in Social Psychology*. Ed. / Trans. Gerard Duveen. New York: New York University.

Niemeyer, Simon and Dryzek, John, 2007. "Intersubjective Rationality." Paper presented at ECPR (European Consortium on Political Research), Helsinki.

Ostrom, Elinor, 1990. *Governing the Commons: The Evolution of Institutions of Collective Action*. Cambridge, UK: Cambridge University Press.

Phillips, John, 2006. "Agencement / Assemblage." *Theory, Culture and Society*, 23 (2–3): 109–109.

Postone, Moishe, 1993. *Time, Labor and Social Domination*. Cambridge, UK: Cambridge University Press.

Roth, Abraham, 2005. "Practical Intersubjectivity," in Frederick F. Schmitt, ed. *Socializing Metaphysics: The Nature of Social Reality*. Latham: Rowman and Littlefield, pp. 167–193.

Rouse, John, 2006. "Practice Theory," in Stephen Turner and Mark Risjord, eds. *Philosophy of Anthropology and Sociology/ vol.15, Handbook of Philosophy of the Sciences*. Dordrecht, North Holland: Elsevier, pp. 639–681.

Sawyer, R. Keith, 2001. "Emergence in Sociology: Contemporary Philosophy of Mind and Some Implications for Sociological Theory." *American Journal of Sociology*, 107 (3): 555–585.

Schatzki, Theodore, 2008. *Social Practices: A Wittgensteinian Approach to Human Activity and the Social*. Cambridge, UK: Cambridge University Press.

Schutz, Alfred, 1967. *The Phenomenology of the Social World*. Evanston: Northwestern University Press.

Searle, John, 1995. *The Construction of Social Reality*. Cambridge, UK: Cambridge University Press.

Steinberg, Alexandra, 2007. "Rhizomic Network Analysis: Toward a Better Understanding of Knowledge Dynamics of Innovation in Business Networks," in Fang Zhao, ed. *Handbook of Research on Information Technology, Entrepreneurship and Innovation*. Hershey: The Idea Group, pp. 224–249. Derived from Steinberg's 3005 Ph. D. dissertation on Emergent Rhizomic Becomings," London School of Economics

Torfing, Jacob, 1999. *Theories of Discourse: Laclau, Mouffe, Zizek*. Oxford: Blackwell.

Touraine, Alain, 1977. *The Self-Production of Society*. Trans. Derek Coltman. Cambridge, UK: Cambridge University Press.

Touraine, Alain, 2009. *Thinking Differently*. Trans. David Macey. Cambridge, UK: Polity Press.

Utting, Peter, (ed.) 2014. *Social and Solidarity Economy: Beyond the Fringe*. London: Zed Books.

Van Wezemael, Joris, 2008. "The Contribution of Assemblage Theory and Minor Politics for Democratic Network Governance." *Planning Theory*, 7 (2): 165–185.

Voss, Jean-Peter and Kemp, René, 2006. "Sustainability and Reflexive Governance: Introduction," in J.P. Voss, Dierck Bauknecht and René Kemp, eds. *Reflexive Governance for Sustainability*. Cheltenham: Edward Elgar, pp. 3–28.

Wolin, Sheldon, 1994. "Fugitive Democracy." *Constellations*, 1 (1): 1–25.

# 2 Sustainable Community Movement Organizations (SCMOs)

## Reinvigorating cooperatives and mutualités via post-capitalistic practices

*Francesca Forno and Paolo R. Graziano*

### Introduction

Contemporary social movement organizations are facing two interrelated problems at the same time. On the one hand, they are confronting institutions that seem less able or willing to accommodate demands for social justice and equity. On the other, they have to deal with the highly individualized nature of contemporary society, which makes difficult to create bonds of solidarity and cooperation among people, the necessary conditions for collective action (D'Alisa et al., 2015; Forno and Graziano, 2014).

Within recent debate on collective action and various forms of activism involved in organizing alternatives to bring about social changes, the concept of Sustainable Community Movement Organizations (hereafter SCMOs) has been suggested to indicate those experiences that try to rebuild social bonds by constructing alternative, productive, and sustainable networks of production, exchange, and consumption (Forno and Graziano, 2014).

Although with some differences due to their countries of origin, SCMOs share several common traits regarding both their motivations and organizational structures. Departing from a similar critique of the contemporary individualized consumerist lifestyle, SCMOs main aim is to support sustainable ways of production and consumption by creating new economic and cultural spaces for civic learning and consumerist actions to sustain alternative markets based on knowledge exchange, loyalty, and trust. In their attempt to do so, SCMOs work to bring different collectives together to help them to develop alternative socio-economic systems in which the overriding object of profit maximization is substituted by cooperation, solidarity, and mutualism. The alternative economic networks set in place facilitate both the circulation of resources (information, tasks, money, and goods) and the construction of common interpretations of reality, thus simultaneously providing a framework for collective action and enabling the actual deployment of alternative lifestyles (Forno et al., 2015). SCMOs also share the idea of abandoning ecologically destructive practices of consumption to favour more sustainable ways of living based on the valorization and revitalization of the local economy.

Although SCMOs are hugely indebted (also in terms of activists) to movements of the past (della Porta, 2007; Forno and Graziano, 2014; Forno and Gunnarson, 2011; Guidi and Andretta, 2015), unlike earlier mobilizations, they are more oriented towards building constructive and thoroughly organized alternatives, rather than appealing to formal institutions by putting pressure so as to make them change their political decisions. Put simply, rather than asking for change, these movement actors aim to produce the change they want to see in the society by building concrete alternatives to the system they are contesting. In order to do so, they act simultaneously on cultural, economic, and political levels.

On a cultural level, SCMOs promote alternative lifestyles and values to oppose consumerism as an economic order that encourages the endless consumption of finite resources in the name of exponential economic growth. Through the organization of activities, such as farmers' markets, conferences, festivals, guided visits to local producers, and a skilful use of new and old media, they are also creating and promoting a "new social imaginary," which implies a particular way of organizing production and consumption and a prioritization of environmental, political, and consumerist values (Latouche, 2010). By addressing basic necessities of everyday life, SCMOs go to the heart of many of the systemic dustuncitons and contradictions that characterize the present model of production, distribution, and consumption. They do so by proposing a new interpretation of the reality based on an alternative symbolic discourse of the social world helping to create new ways of living together and new ways of representing collective life (Thomson, 1984). Giving importance to economy relations based on sharing, gifts, and conviviality, is fundamental in order to generate, sustain, and spread alternative economic projects.

On an economic level, these experiences encourage greater economic self-sufficiency and facilitate the construction and sustainability of alternative economic circuits, which favour services and products that respect certain ethical standards, such as fair trade and recycled goods, and the consumption of local, seasonal, fresh, traditional, and often organic produce. Attention is also given to supporting supplies from renewable energy sources to reduce reliance on fossil fuels (insulation, efficient appliances, carpooling, and community transport).

On a political level, through the promotion of political consumerism and sustainable lifestyles (Micheletti, 2003, 2009), SCMOs develop and sustain innovative models of environmental regulatory governance based on voluntary actions and participation, which, instead of imposing a certain kind of behaviour, aim to support and promote more sustainable developments in practice. Research has shown that being a part of such organization can also encourage the diffusion of more collaborative attitudes, fostering interest in politics among participants and enhancing members' sense of social effectiveness (Forno et al., 2015, p. 83). Furthermore, setting bridges between local consumer practices and local representative politics, SCMOs help their members to develop new civic awareness and democratic competences. In some cases, SCMOs have also turned into sorts of lobbying organizations, and in some cases they have even supported civil electoral lists participating in local elections (Graziano and Forno, 2012).

Thus, even though such movement organizations do not mobilize and structure their claims primarily through contentious activities, to a certain extent the contentious dimension of these networks can be seen as embedded in their social and economic networking activities. In fact, it is through solidarity exchanges that these organizations support strategies of direct action (Bosi and Zamponi, 2018), such as information sharing, awareness raising, educating, and lobbying.

In other words, SCMOs try to go beyond capitalist settings by encouraging ongoing and direct relationships among different actors (workers, producers, and consumers) based on solidarity and reciprocity rather than economic convenience (i.e., utility or profit maximization). For example, sustained community agriculture organizations create and consolidate local social relationships between producers and consumers, which are also characterized by the presence of a monetary exchange (i.e., buying specific products) but is primarily centred on a social relationship – not on a commercial one. Put differently, within the networks created by SCMOs, the commercial or economic exchange are a by-product of a social exchange (relationship) and not vice versa (Forno and Graziano, 2014; Graziano and Forno, 2012).

## The culture and organization of SCMOs

In cultural terms, the expansion of SCMOs comes both from an alternative political culture and from the rising concern about the consequences of economic policies placing emphasis on unlimited growth of consumption and production. Although SCMOs might emphasize different aspects and act on different scales of action (see below), they do share a common understanding of how society works and how people are influenced in their decision-making behaviour by their social context.

First, the prevailing cultural traits move from a strong criticism towards materialism and standard consumerism. At this regard, SCMOs point out that current standards of consumption and production damage the environment, contribute to climate change and to use up resources at a rate which is unsustainable. Furthermore, SCMOs implicitly (and often explicitly) denounce the excessive attention paid only to the price of products which has undermined the guarantee of labour standards and accentuated exploitation of workers with the aim to reduce the overall unitary cost of labour, a specific cultural trait that, as argued below, derives from the *Global Justice Movement's* (GJM) focus on environmental social justice concerns (Andretta et al., 2002).

Second, SCMOs are particularly interested in contrasting mass production and supporting artisanal products, natural materials, and handmade items. The general attitude is not necessarily a "luddist" one since only few "anti-consumerism" movements (such as, at least to a certain extent, the de-growth movement) are against production *per se*. The general attitude is, rather, an attitude which looks favorably to small-scale production guaranteeing fair profits for the workers and limits (if not abolishes) the retribution of intermediaries in the value chain. This is, for example, the case of fair trade organizations that provide a social premium to producers – mainly small cooperatives – in order for them to benefit from set prices (typically above market prices) and better and fairer allocation and distribution of the profits.

Third, SCMOs do focus primarily on the local level, for example, stressing the importance in daily consumption to prefer local food and products coming from local business and local economies. At this regard, it is however important to note that the local dimensions emphasized by SCMOs are not "parochial" but rather "cosmopolitan-local." As it will be argued also below, SCMOs rose in fact along the lines traced by those organizations – associations, cooperatives, small- and medium-sized enterprises, etc. – that during the *GJM* have actively contributed to translate conscious consumerism and grassroots advocacy into political action by stimulating both the production and commercialization of environment-friendly and fair trade products and a new global consumer culture. Therefore, SCMOs, while supporting local producers (and/or community projects), are also concerned with the transnational distribution of wealth and life opportunities. Namely, although their focus is primarily on the local level, within these organizations "the local is not usually idealized as a space insulated from power relations and anomic global capitalism but is acknowledged as a publicly contested site of political economic struggle, exploitation and accumulation" (Goodman et al., 2012, p. 8).

For example, the several local food movements that have spread over the past decade all over the world promote local products but at the same time support transnational food networks in order to increase food cultures and experiences (as in the case of the biennial event of *Terra Madre* promoted by Slow Food International, where local producers from all over the world meet in order to present to a wide – paying – audience their products and food cultures). Such example shows how peculiar the global/local link can be in the cases of SCMOs, Slow Food being a case where the predominant local activity is inserted into a global framework of action.

Finally, another specific trait of SCMOs culture is the presence of diffused mutual solidarity – not only between producers and consumers but also among consumers and among producers. As mentioned above, relationships and commitments go beyond the existence of a commercial or economic exchange and in case of specific needs reciprocity between actors is supported by active cooperation between producers and consumers (for example under the form of low-interest rate loans) or among consumers (for example, delivery costs may be shared) and among producers (for example, by supporting the creation of producers' links in order to support "solidarity economic districts"). Such mutual solidarity is deeply rooted in the territory but is often facilitated by new technologies, such as the internet, which make communication and connections faster and cheaper.

Acting locally by thinking globally is a slogan which is very well-suited for SCMOs, although – as it will be discussed in the following sections – the mix between global and local targets vary to a great extent, as the specific attitude of alter- or anti-consumerism expressed by the organizations.

## Re-embedding the economy into society for a sustainable world

The shift from the global to the national and local level observed within SCMOs can be read as a consequence of the limited opportunities offered by the external context encountered by GJM – the global network of social movement organizations opposing for the first time to what is often known as the "corporate globalization," which emerged at the end of the 1990s.

As it is often underlined, social movements mobilization capacity is conditioned by the characteristics of the environment within which grassroots actors find themselves to operate. For example, after the early 2000s the GJM did not seem to be able to exert a great influence on transnational decision-making, and this was particularly due to the lack of opportunities provided by the international governance system.

The lack of institutional allies, the internal and problematic differentiation among the various components of the movement, the violence with which certain demonstrations were repressed, and the disappointment of activists for the negative outcome of the great popular mobilization against the war in Iraq in 2003 led to a rapid decline of the movement. However, the end of the cycle did not mean the end of collective action but gave birth to a substantial re-positioning of some social movement organizations from the global to the local scale of action.

In fact, it is at the local level that grassroots activism has continued to spread and expand by virtue of the emergence and consolidation of several new initiatives aiming at raising awareness regarding sustainable living, ecological preservation, and greater respect of workers' and human rights. Examples of such efforts include barter groups, urban gardening collectives, new consumer-producer networks and cooperatives, recovered factories, local savings groups, time banks, urban squatting, and other similar experiences.

As argued by Melucci (1996), networks and activities do in fact persist even when movements are not publicly mobilizing. According to his latency-visibility model (Melucci, 1985), it is moreover during the phase of 'latency' that new meanings and identities are often forged and new repertoires of action are put to test. As empirical research has started to show (Forno, 2015; Grasseni, 2013), in the formation of SCMOs local friendship and kinship ties clearly play an important role. In many cases, however, what also appeared to be important is the common experience among "core activists" (or local initiators) in groups and organizations that had participated in the *GJM*. For example, within SCMOs it is common to find people that during the *GJM* protest cycle had been active in movement organizations, such as *Via Campesina*, the family farmers' international; *Peoples' Global Action*, a loose collection of often youth groups; *Jubilee 2000*, the Christian-based movement for relieving international debt; *Friends of the Earth*, the environmentalist international; and some think-tanks, like *Focus on the Global South* and *Third World Network*, as well as some large and transnational trade union organizations, trade unions, faith-based and peace groups, etc.

More in general, as it has been often observed, social movements tend to emerge and expand in situations where the political, economic, and social opportunities are neither fully open nor completely closed to these actors' claims. That means that mobilization tends to remain confined within small groups of activists when the instances "from below" are either immediately implemented and channelled through traditional interest mediation routes, or fail to gain support from powerful allies and/or are repressed through violence (Kriesi et al., 1995). The degree of opening and closing of the opportunities systems towards the instances put forward by social movement actors is also important with regard to both their repertoire of action and their organizational structure.

When looking at the history of social movements, for example, it is possible to recognize some historical periods during which they have directly opposed the dominant institutional powers, and others in which movements have challenged the structure of power by proposing and supporting forms of self-organization. This is the case for example of the movements that have emerged and developed during the first half of the 19th century, such as the mutualistic and co-operative movements (Forno, 2013).

The history of the cooperatives and mutualités is particularly important to understand the shift observed in several social movement organizations from the global to the local level. During the years of the industrial revolution – an historical time characterized by great changes and a system of opportunities not particularly favourable to the action of social movements – the co-operatives became an instrument of economic organization and emancipation for workers (Hilson, 2009, 2011; Scott, 1998). Regardless of their type, size, geographical location, or purpose, cooperatives provided a tool through which it was possible to achieve one or more economic goals, such as improving bargaining power when dealing with other businesses, bulk purchasing to guarantee lower prices, obtaining products or services otherwise unavailable, gaining market access or broadening market opportunities, improving product or service quality, securing credit from financial institutions, and increasing income (Forno, 2013). As noted by Mayer, to a certain extent, consumer cooperatives, particularly food cooperatives, have been rather important for reducing the cost of goods and for empowering citizens by giving them more control over economic institutions at the community level (Mayer, 1989: 81).

Moreover, through their educational work, cooperatives were also fundamental for the spread of a culture of cooperation and the formation of social ties among their members, which in several cases led to collective actions undertaken by specific social groups (Gurney, 1996).

Although today we are clearly in a different situation, many traits that characterize SCMOs seem to recall forms of self-organization of the past. As it was for cooperatives and mutualités, SCMOs too address both the intensification of economic problems and the difficulties of rebuilding social bonds within society, emphasizing solidarity and the use of "alternative" forms of consumption as means to re-embed the economic system within social relations, starting from the local level. After all, an interesting common trait that distinguishes SCMOs from other contemporary social movement organizations is the emphasis simultaneously given to both consumption and production.

In other words, faced with increasing environmental and financial challenges, new forms of cooperation between consumer and producer groups are somehow regaining momentum also beyond Europe and North America, to some extent reappraising the 19th-century ideas on the need to reorganize economic life on the basis of human and social needs, beyond (and against) capitalist accumulation. Similarly with the past, through SCMOs individuals have the opportunity not only to satisfy a series of consumer-related needs in an ethical way, but also to join together to make their voices heard (primarily on environmental and social justice issues).

With regard to emergence and diffusion of SCMOs, particularly important during the 20th century was the experience of the Fair Trade movement which aimed to help producers in developing countries to make better trading conditions and promote sustainability.

The convergence of people coming from different experiences is particularly interesting in the context of SCMOs. Moreover, by concentrating their action also in the creation of alternative provisioning systems (as, for example, alternative food markets where organic products can be purchased at a lower price), SCMOs often attract individuals who may also at the first place just be interested in purchasing good quality products at a lower price. Such people, by simply purchasing, would learn that a certain type of consumption means also better land management, support to small producers, and greater attention to the production process not only from the technical point of view but also for those who work there. This means, for example, that the organizations that support fair trade, eco-tourism, etc., have increasingly become a mix of hardcore activists and other consumers which have been attracted by the availability of convenient products and not motivated "politically."

The rise and spread of grassroots initiatives aiming at building alternative and sustainable networks of production, exchange, and consumption has continued also during the years of the economic crisis (Bosi and Zamponi, 2015; Castells et al., 2012; D'Alisa et al., 2015; Forno and Graziano, 2019; Giugni and Grasso, 2018; Koos et al., 2017; Kousis, 2017; Lekakis, 2017). The 2007–2008 global financial crisis seems in fact to have further fuelled the emergence of more locally based social movement actors in some cases in conjunction with the Indignados (aka the 15M movement) and *Occupy Wall Street* mobilization.

## A typology of SCMOs

Beside similarities, SCMOs do also present important differences. Using two key dimensions of differentiation – scale of action and attitude towards consumption – Forno and Graziano (2014) proposed an inductive classification of the various types of organizations engaged in the construction of discourses and alternative economy projects. Table 2.1 summarizes the results of such classification, giving evidence of a variety of SCMOs that have grown over recent decades, creating spaces for political consumerist actions. Of course, it is important to note that these are not mutually exclusive categories, in that the same SCMO could work on a different scale of action as well as adopting or promoting both alter- and anticonsumerist practices and discourses.

*Table 2.1* Types of SCMOs

| *Predominant attitude towards consumption* | |
|---|---|
| *Alter-consumerisms* | *Anti-consumerism* |
| **Predominant scale of action** | |
| *Global* | |
| Fair trade<br>Ethical Fashion (no-sweat groups) | Group promoting de-growth<br>Simplicity movement<br>*Buen vivir* |
| *Local* | |
| Farmers' market<br>Slow Food movement<br>Community-supported agriculture (CSA)<br>Rotating Credit and Savings Associations (ROSCAs) | Transition towns<br>Intentional communities (e.g. Ecovillages, cohousing communities) |

Source: Adapted from Forno and Graziano, 2014.

The first possible type of SCMOs is represented by an alterconsumerist attitude and a predominant global scale of action. This is the case, for example, of trading partnership, based on dialogue, transparency, and respect, that seeks greater equity in international trade (Goodman et al., 2012; Raynolds et al., 2007). Examples are fair trade organizations and other alternative economic networks dedicated to improving working conditions and supporting the empowerment of workers in global industries (Balsiger, 2014), on the one hand, through lobbying, awareness raising and, on the other, by creating new economic and cultural spaces for the trading, production, and consumption of goods whose ethical and aesthetic alternative qualifications distinguish them for products conventionally supplied by international trade, mainstream food manufacturers, and supermarket chains (Goodman et al., 2012).

Another possible type is represented by those SCMOs that have an alterconsumerist attitude and act predominantly on a local scale. Key examples here are all those alternative economic networks such as barter groups, time banks, local savings groups/alternative currencies (see Varvarousis et al. and Hossein, this volume) that have grown intensely over recent decades. As argued, such experiences represent a grassroots response to the global financial crisis and have virtually mushroomed since the turn of the millennium, in the North and in the South, in urban and rural contexts.

Particularly preeminent are those SCMOs that have emerged in the field of food. Organizations such as solidarity purchase groups (see Dal Gobbo and Forno, this volume), community-supported agriculture, Slow Food, and new consumer-producer cooperatives aim to bring about a process of relocalization and resocialization of food production-distribution-consumption practices, with a view to the construction of a more environmentally sound, socially just, and economically sustainable local food system (Brunori et al., 2011; Dubuisson-Quellier and Lamine, 2008; Goodman et al., 2012).

A third type, characterized by an anticonsumerist attitude and a local scale of action, is represented by all those SCMOs promoting community projects that aim to increase self-sufficiency to reduce the potential effects of peak oil, climate destruction, and economic instability. Some examples are intentional communities, such as ecovillages, and transition towns. People who create or move into such experiences see consumerist lifestyles, the breakdown of traditional forms of community, damage to the environment, and human overreliance on fossil fuels as trends that need to be changed to prevent social and ecological disaster. While living in ecovillages entails living in a community, the term "transition town" refers to grassroots local group initiatives that aim to increase self-sufficiency to reduce the potential effects of energy depletion, climate destruction, and economic instability (Sage, 2014).

Finally, a fourth type, characterized by an anticonsumerist attitude and a global scale of action, includes those movement organizations promoting a radical view that draws inspiration from deep ecology. Examples are groups promoting voluntary simplicity and de-growth in the Global North and the *buen vivir* movement in the Global South (especially across South America: see Neyra in this volume). Although diverse in many ways, all these SCMOs, which have also significantly gained in popularity over the past decades, claim that saving the environment will require radical changes in how our society functions and interacts with the planet.

Voluntary simplicity means taking the decision to live more simply, to consume less and better, and to work less and devote more life energy to relationships and cultural activities (Alexander and Ussher, 2012). While the focus on sustainable economic practices – sharing work and consuming less while devoting more time to art, music, family, nature, culture, and the community – is also central to the perspective of de-growth activism, this movement presents itself as a radical political, economic, and social project based on ecological economics and anticonsumerist and anticapitalist ideas (Asara et al., 2015; D'Alisa et al., 2014; De Maria et al., 2013). As a worldview, *buen vivir* raises from a cosmovision that encompasses all the aspects of life. From this perspective it is claimed that humans are understood not as owners, but as reciprocating stewards of the Earth and its resources. Nature is therefore considered not as "natural capital" but as a being without which life does not exist.

Contrary to some expectations about political consumerism and producerism (Andretta and Guidi, 2017) being closely associated with higher levels of wealth, the current time characterized by a "multiple crisis" seems to show that SCMOs can be given further impetus by austerity concerns faced by individuals and communities (Conill et al., 2012; D'Alisa et al., 2015; Guidi and Andretta, 2015; Kousis, 2017). Buying food directly from producers, going to local markets or swapping food in an urban warehouse, living in an ecovillage, or taking part in an informal financial institutions are indeed practices that can be now observed worldwide as a global phenomenon.

## Conclusions

SCMOs represent life-style alternatives and forms of resistance to the traditional marketplace (Papaoikonomou and Alarcon, 2017). SCMOs try in fact to go beyond capitalist settings by encouraging ongoing and direct relationships between different actors (workers, producers, and consumers) based on solidarity and reciprocity rather than economic convenience (i.e., utility or profit maximization).

Along with a series of studies underlying the innovative character of SCMOs, economic activism has also faced a number of criticisms. In particular, it was argued that by focussing on self- determination and self-changing strategies, these experiences divert civic action from real economic and social problems, promoting political-ideological formulas that can channel social discontent away from the real targets, such as national and international institutions (Rosol, 2012). Similarly, other critics have stressed the limits of these movements in terms of their transformative potential and efficacy. For instance, Goodman et al. (2012), while discussing the case of Alternative Food Networks (AFNs), argue that these experiences are often the expression of the middle and upper classes, which are often scarcely politicized and more interested in preserving their own health and identity. From this observation follows the claim that community-organized collective consumption is bound to remain a niche on the edge of the market, with limited or no impact on how society functions, and that it is easily co-opted by corporate marketing strategies (Dubuisson- Quellier, 2019).

With regard to these criticisms, empirical studies have showed how usually those engaged in such organizations often also tend to be active in other forms of participation. For instance, among members of the Italian Solidarity Purchase Groups, various studies have found that many of them have had previous participatory experiences in other social movement or voluntary organizations (Forno et al., 2015; Guidi and Andretta, 2015). Moreover, the political activities of SCMOs often include mobilization against environmental, economic, and urban policies at a global and local level, along with environmental organizations, grassroots associations, and trade unions (Guidi and Andretta, 2015). As Schlosberg and Coles (2015) argued,

> food movements in the United States, United Kingdom and Australia aim to build new circulations of localist food economy, but continue to lobby and protest for changes in state, national and transnational food policy as well. Again, one form of political engagement does not simply replace another; new materialist political action is not a zero-sum or an either/ or.

Finally, while on the consumption end these experiences often indeed look like an expression of the middle and upper classes, this can be radically different on the production end, where there are often small-scale producers at risk of being excluded from the market. By joining people – who may even initially hold different and possibly even conflicting views about how society and its economic system should actually work in order to become more sustainable and fairer – starting from material and often daily needs, these networks offer new spaces of apprenticeship for a new type of consumer citizenship (Grasseni, 2013, 2014). Overall, in our individualized and fragmented societies, these groups represent an important way for people to bond together, building social capital and multiplying their potential to act.

Of course, these experiences are not all the same, and some networks may be more successful and more long-lasting than others. Differences may depend either on how the various initiatives are constrained by and in turn affect the polities and economies within which they are embedded, on the form and sustainability of their organizations as well as on external or preexisting organizations, and also on activists' motivation to act and their capacity to mediate between instrumental and noninstrumental rationales.

The 2007–2008 crisis seems to have given a further impetus to the spread of these experiences, confirming the intuition of Castells et al. (2012) that the alternative economic sector is one of the four emerging layers of EU and USA economies after 2008. In fact, beyond promoting and practicing solidarity market exchanges, alternative economic networks are often incubators and accelerators of innovative local economy initiatives. And this seems increasingly true in both the global North and in the global South.

## Bibliography

Alexander, S. and Ussher, S. (2012) The voluntary simplicity movement: A multi-national survey analysis in theoretical context. *Journal of Consumer Culture*, 12(1), 66–84.

Andretta, Massimiliano and Guidi, Riccardo (2017) "Political Consumerism and Producerism in Times of Crisis. A Social Movement Perspective?" *Partecipazione & Conflitto*, 10(1), 246–274.

Andretta, M., della Porta, D., Mosca, L. and Reiter, H. (2002) *Global, Noglobal, New Global, La protesta contro il G8 a Genova*. Bari: Laterza.

Asara, V., Otero, I., Demaria, F. and Corbera, E. (2015) Socially sustainable degrowth as a social–ecological transformation: Repoliticizing sustainability. *Sustainability Science*, 10(3), 375–384.

Balsiger, P. (2014) Between shaming corporations and promoting alternatives: An in depth analysis of the tactical repertoire of a campaign for ethical fashion in Switzerland. *Journal of Consumer Culture*, 14(2), 218–235.

Bosi, L. and Zamponi, L. (2018) Political consumerism and participation in times of crisis in Italy. In: Giugni, M. and Grasso, M. T. (eds), *Citizens and the Crisis*. London: Palgrave, pp. 141–164.

Brunori, G., Rossi, A. and Malandrin, V. (2011) Co-producing transition: Innovation processes in farms adhering to solidarity- based purchase groups (GAS) in Tuscany, Italy. *International Journal of Sociology of Agriculture and Food*, 18(1), 28–53.

Castells, M., Caraca, J. E. and Cardoso, G. (ed) (2012) *Aftermath. The Cultures of the Economic Crisis*. Oxford: Oxford University Press.

Collins, J. (2003) *Threads: Gender, Labor, and Power in the Global Apparels Industry*. Chicago, IL: University of Chicago Press.

Conill, J., Castells, M., Cardenas, A. and Servon, L. (2012) Beyond the crisis: The emergence of alternative economic practices. In: Castells, M., Caraça, J., and Cardoso, G. (eds), *Aftermath: The cultures of the economic crisis*. Oxford: Oxford University Press, pp. 210– 250.

D'Alisa, G., Demaria, F. and Kallis, G. (eds) (2014) *Degrowth: A Vocabulary for a New Era*. London: Routledge.

D'Alisa, G., Forno, F. and Maurano, S. (2015) Grassroots (economic) activism in times of crisis: Mapping the redundancy of collective actions. *Partecipazione and Conflitto*, 8(2), 328–342.

della Porta, D. (ed) (2007) *The Global Justice Movement: Cross-national and Transnational Perspectives*. Boulder, CO: Paradigm Publishers.

Demaria, F., Schneider, F., Sekulova, F. and Martinez-Alier, J. (2013) What is degrowth? From an activist slogan to a social movement. *Environmental Values*, 22, 191–215.

Dubuisson-Quellier, S. and Lamine, C. (2008) Consumer involvement in fair trade and local food systems: Delegation and empowerment regimes. *GeoJournal*, 73(1), 55–65.

Dubuisson-Quellier, S. (2019) From moral concerns to market values: How political consumerism shapes markets. In: Boström, M., Micheletti, M. and Oosterveer, P. (eds), *The Oxford Handbook of Political Consumerism*. Oxford: Oxford University Press, pp. 813–832.

Forno, F. (2013) Co-operative movement. In: Snow, D. A., Della Porta, D., Klandermans, B. and McAdam, D. (eds), *Blackwell Encyclopedia of Social and Political Movements*. Oxford: Blackwell Publishing, vol. I, pp. 278–280.

Forno, F. (2015) Bringing together scattered and localised actors: Political consumerism as a tool for self-organizing anti-mafia communities. *International Consumer Studies*, 39(5), 535–543.

Forno, F. and Graziano, P. R. (2014) Sustainable community movement organisations. *Journal of Consumer Culture*, 14(2), 139–157.

Forno, F. and Graziano, P. R. (2019) From global to glocal. Sustainable Community Movement Organisations (SCMOs) in times of crisis. *European Societies*, 21(5), 729–752.

Forno, F. and Gunnarson, C. (2011). Everyday shopping to fight the Mafia in Italy. In: Micheletti, M. and McFarland, A. S. (eds), *Creative participation: Responsibility- taking in the political world*. London and Boulder, CO: Paradigm, pp. 103–126.

Forno F., Grasseni C. and Signori, S. (2015) Italy's solidarity purchase groups as "citizenship labs." In: Huddart Kennedy, E., Cohen, M. J. and Krogman, N. (eds), *Putting Sustainability into Practice: Advances and Applications of Social Practice Theories*. Cheltenham: Edward Elgar, pp. 67–88.

Giugni, M. and Grasso, M. T. (eds) (2018) *Citizens and the Crisis*. London: Palgrave.

Goodman, D., DuPuis, E. M. and Goodman, M. K. (2012) *Alternative Food Networks: Knowledge, Practice, and Politics*. London: Routledge.

Grasseni, C. (2013) *Beyond Alternative Food Networks: Italy's Solidarity Purchase Groups*. London: Berg/Bloomsbury Academic.

Grasseni, C. (2014) Seeds of trust: Italy's gruppi di acquisto solidale (solidarity purchase Groups). *Journal of Political Ecology*, 21(1), 179– 192.

Graziano, P. R. and Forno, F. (2012) Political consumerism and new forms of political participation: The Gruppi di Acquisto Solidale in Italy. *Annals AAPSS*, 644, 121–133.

Gurney, P. (1996) *Co-operative Culture and the Politics of Consumption in England, 1870–1930,* Manchester: Manchester University Press.

Guidi, R. and Andretta, M. (2015) Between resistance and resilience: How do Italian solidarity based purchase groups change in times of crisis and austerity? *Partecipazione e confitto*, 8(2), 443–447.

Hilson, M. (2009) The consumer co-operative movement in cross-national perspective: Britain and Sweden, c. 1960–1939. In Black L. and Robertson N. (eds), *Consumerism and the Cooperative Movement in Modern British History*, Manchester: Manchester University Press, pp. 1918–39.

Hilson, M. (2011) A consumers' international? The international cooperative alliance and cooperative internationalism, 1918–1939: A Nordic perspective. *International Review of Social History*, 52(221), 1–31.

Koos, S., Keller M. and Vihalemm T. (2017) Coping with crises: Consumption and social resilience on markets. *International Journal of Consumption Studies*, 41(4), 363–370.

Kousis, M. (2017) Alternative forms of resilience confronting hard economic times: A South European perspective. *Partecipazione and Conflitto*, 10(1), 119–135.

Kriesi, H., Koopmans, J. W. and Giugni, M. (1995) *New Social Movements in Western Europe*. London: UCL Press.

Latouche, S. (2010) *Farewell to Growth*. London: Polity Press.

Lekakis, E. J. (2017) Economic nationalism and the cultural politics of consumption under austerity: The rise of ethnocentric consumption in Greece. *Journal of Consumer Culture*, 17(2), 286–302.

Mayer, R. N. (1989) *The Consumer Movement: Guardians of the Marketplace*. Boston, MA: Twayne.

Melucci, A. (1985) The symbolic challenge of contemporary movements. *Social Research*, 52(4), 789–816.

Melucci, A. (1996) *Challenging Codes: Collective Action in the Information Age*. Cambridge: Cambridge University Press.

Micheletti, M. (2003) *Political Virtue and Shopping. Individuals, Consumerism and Collective Action*. London: Palgrave Macmillan.

Micheletti, M. (2009) La svolta dei consumatori nella responsabilità e nella cittadinanza. *Partecipazione e Conflitto*, N3, 17–41.
Papaoikonomou, E. and Alarcón, A. (2017) Revisiting consumer empowerment: An exploration of ethical consumption communities. *Journal of Macromarketing*, 37(1), 40–56.
Raynolds, L. T., Murray, D. and Wilkinson, J. (2007) *Fair Trade, the Challenges of Transforming Globalization*. New York: Routledge.
Rosol, M. (2012) Community volunteering as neoliberal strategy? Green space production in Berlin. *Antipode*, 44(1), 239–257.
Sage, C. (2014) The transition movement and food sovereignty: From local resilience to global engagement in food system transformation. *Journal of Consumer Culture*, 14(2), 254–275.
Schlosberg, D. and Coles, R. (2015). The new environmentalism of everyday life: Sustainability, material flows, and movements. *Contemporary Political Theory*, 15(2), 160–181.
Scott, G. (1998) *Feminism and the Politics of Working Women: The Women's Co-operative Guild, 1880 to the Second World War*, London: Routledge.
Thomson, J. B. (1984) *Studies in the Theory of Ideology*. Berkeley: University of California Press.

# 3 Operationalizing SCMO as a social-economic concept

## The case of post-2008 Spain

*Richard R. Weiner and Iván López*

The sustainable community movement organization (SCMO) is a new mediating form of social solidarity and social space in social-economics that goes beyond our traditional concept of mutual aid, cooperatives and mutualités. It is a form that tries to alter the conventions of contemporary market practices with social entrepreneurship practices. Its distinguishing characteristic is its networking evolution of co-responsibility in platform-based reciprocal solidarity to generate sustainable ecological conditions necessary for socially self-managed development.

Linked to the social-economics of Elinor Ostrom (1990), the co-responsibility and stewardship embodied by SCMOs involve pooling what were once seen as purely private goods in an inter-connectedness of reciprocal solidarity. Ostrom focuses on the rules and institutions governing "the commons." This chapter is less a rational choice institutionalist approach, and more one that focuses on the process of commoning in terms of solidarity, spatiality and support. SCMO meshworks of endogenous trust generate ecological conditions necessary for socially self-managed development, a new form of the 19th-century concept of *mutualité* and cooperatives. And as we begin, we should recall this quote by Marx from 1850.

> Only when the real human individual reincorporates in itself the abstract citizen of the State, when he or she as an individual, becomes at the same time a generic human in his or her empirical life, in his or her individual work, in his or her individual condition; and only when he or she recognizes his or her strength from within (forces propres) are social forces and organizes them accordingly, no longer separating himself or herself from the power of society in the form of political power – Only then will human emancipation be complete.
>
> (Karl Marx, 1850 "Our System, or the Self-Negating Worldly Wisdom and the Globalizing Movement of Our Time")[1]

## Introduction: SCMO from a social-economic perspective

SCMO strategy is to act on both sides of the producer/consumer divide, getting consumers to keep in mind the material needs of producers, while getting producers to consider environmental externalities of production and the inclusion of consumers in production decision-making. In the process, the SCMO strategy aims at re-embedding the economic within the social.

Further, SCMO strategy includes what Forno and Graziano (2014) label "political consumerism" – an application of what Albert Hirschman (1970) labeled "voice" options of demonstrations and "exit" options of alternative autonomous practices outside of the economic mainstream, especially at the local level. SCMOs not only change lifestyles toward more sustainable forms of consumption and production, it develops a collective conscience/consciousness of SCMO members' social capital, as in engaging in the wake of the de-alignment of party politics since the 2008 Crisis.

SCMOs have been somewhat successful in helping to set standards for environmental regulatory governance as a political consumerism of protocolism involving collective responsibility of both producers and consumers, rather than statist coercive imposition of rules. Importantly, as Forno and Graziano, 2016, 6) note:

1 SCMOs are characterized by a sense of diffused mutual solidarity;
2 "Such mutual solidarity is deeply rooted in the (local) territory but is often facilitated by new technologies, such as the Internet, which makes communications and connections faster and cheaper";
3 And while SCMOs' focus is primarily on the local level, they have shown attentiveness to both guaranteed labor standards and transnational effects on the distribution of wealth and life chances.

Recalling the prescient appreciation of the coming precarity of young workers by André Gorz (1977/1980), we can understand SCMOs as part of both Solidarity Economy and Degrowth movements. Both movements share a sensibility of a need for a river going back to its normal flow after a disastrous flood, as in unbridled growth and capital accumulation. An emergent degrowth movement trajectory emphasizes building interconnecting networks of ecological resilience, focusing on a curb on consumption, environmental extraction and waste. Social and Solidarity Economies (SSE) act as enveloping core innovating forms of life – ones constituted by resilient stakeholder mutual recognition and cooperation in the conscious control of the means of production and consumption.

Social self-management means that mutually involved stakeholders have to live with the decisions they make and their consequences – *vis à vis* the society, employment and the environment with which they are interconnected (cf. Monticelli, 2018). Significantly, solidarity economy and social economics eschew collectivist organization and centralized planning. SCMOs emerge in light of the lack of capability of State-run institutions to address economic and environmental crises. SSE encourages social enterprise with a concern for ecology, inclusion, a sense of job security and employment for the poor.

A SCMO involves embodied co-responsibility and stewardship for pooling what were once called private goods, now as common goods. It involves co-producing stakeholders driving sustainable value networks. The economic exchange process is understood as a by-product of a social exchange relationship, not *vice versa* (Forno and Graziano, 2014, 6), focused on the valorization and revitalization of the local economy. And the constitutive provenance of these SCMOs lies in heterarchical multi-stakeholder social pacts. SCMOs try to go beyond "the varieties of capitalism" approach.

What is constituted in practice is what Charles Tilly (2008) called a *trust regime* or a *trust network* wherein an imagination manifests itself for re-assembling institutional practices. Trust networking is increasingly called paranodality – involving more than a single dominating code. Network nodes are decentered, heterarchically/confederatively rather than centrally coordinated. As local and horizontal collaborative embodied engagements for solidarity-based exchange and consumption, these social pacts intend a radical revision of the market by disrupting assimilated neoliberal understandings of how we consume goods and services.

SCMOs focus on both the creation of "commons" and *commoning* of shareable resources enabled by both collaborative culture and infrastructure of production and exchange as well as distributed/distributive digital networking.

- *Commoning* fosters a collective/cooperative process of social learning and participative decision-making;
- *Commoning* involves founding and enforcing institutions for governing knowledge and resources over time through actions, producing and reproducing connective structures and social bonds over time through actions;

- The SCMOs are the bedrock of the SSE, building value chains and supply chains;
- keeping the "value" produced as much as possible inside the SSE network, as opposed to the private and public sectors of the economy; and
- improving bargaining power with the private and public sectors of the economy.

Concretely, the SCMO involves a new mode of resilient social pact-ing is grounded in the decentered *mutual stakeholder social pact* (MSSPs) – known as the frame agreement (see Karin Bäckstrand, 2006). This is a protocolism of standard-setting and task reciprocity based on continued negotiated rule-making and rule-enforcing. The underlying concept here comes from video compression technology wherein an image is established as a base, and subsequent images are stored only as changes from the base. The intention of such new multi-stakeholder social pacts is to rebuild new social relationships around some radical revision of the market and practices of an excessively commodified environment.

SCMOs are institutionalized by MSSPs locally, and beyond in levels of multi-scalarity. The SCMOs' purpose is the creation of *value chains* of shareable resources embedded within the system of capitalism: such as knowledge, services, goods, labor and solidarity purchasing power. Within this value chain network are co-stakeholders, co-producers, co-owners, co-users and co-responsible *reciproqueteurs*. They seek to improve market access and market opportunities.

This emergent new social-economic form is an autonomous normative ordering, a mutually regulating social-economic form whose constitutive provenance lies in heterarchical multi-stakeholder social pact-ing. As such, it is both an embedded social insertion and an embodied responsibility of pooling common resources. In this manner, they are building along the theoretical lines detailed by Elinor Ostrom (1990) toward an inter-connectedness of reciprocal solidarity and endogenous trust for common resource stewardship. SCMOs are motivated by commitment to the creation of shareable resources along with the democratic governance of such resources to sustain people and the planet. They use markets and their limits – rather than streets – as the battlefields. But beyond the governmentality of neoclassical and neoliberal market-mindedness, SCMOs move to the creation of what Ostrom refers to as *non-rivalrous common pool resources*, both alongside and outside the market.

Beyond the commitment to commoning in value chain networking and governance, SCMOs and SSEs involve what Jean-Luc Nancy (1996) refers to a sense of a "a being with/a being in common." This is a sense of a being that is attentive to differences, attentive – as Paul Ricouer (2000) would say – "to the Other like us but not us." Social solidarity affirms human interconnectedness, while at the same time developing each person to assume responsibilities for the Other in what Donna Massey (1995, 2004) refers to as unboundable and ever-changing *geographies of responsibility*. Such a phenomenon is anticipated in the writings of the young Lewis Mumford: *Sticks and Stones* (1924) and *The Golden Day* (1926). These SCMOs stimulate new awareness as well as new paradigms of choice and responsibility.

In the words of the 1987 Bruntland United Nations Commission Report on Environment and Development, these SCMOs meet "the needs of the present without compromising the ability of future generations to meet their own needs." The SCMOs attempt to appropriately gauge contextual scale – "moving to scale" – in best managing production and consumption. These are not necessarily anti-consumerist but are (1) focused on alternative forms of consumption and (2) multi-scale oriented, neither shunning global context nor just focusing on local scale.

Sustainable community movement organization connotes:

- emphasizing not only price and product quality but what can be called "political consumer behavior"; specifically, how we understand the act of consuming a fundamental part of the production process;
- experimenting beyond political consumerist awareness, in alternative modes of sustainable living;
- co-producing of social reassertion/insertion in mutually referent networks based on MSSPs on a local level;
- constituting eco-systems evolving toward sustainability as an open participatory process of empowered voluntary stakeholders;
- cultivating a moral economy wherein durable social bonds of reciprocity and cooperation can be built in terms of "shared stewardship," re-embedding economics activities in social relations and unveiling a sense of connexion; and
- appropriately gauging contextual scale – "moving to scale" – best managing production and consumption: not necessarily anti-consumerist but: (i) focused on alternative forms of consumption and (ii) oriented in a multi-scalar sense, neither shunning global context nor just focusing on local scale.

For the political sociologist Manuel Castells (2012), networks of SCMOs can be thought of as relational space of rhizomatic practices, positioned horizontally. (Indeed, Castells uses the metaphor "rhizomatic" here in *Networks of Outrage and Hope*, pp. 110–155.) This space of rhizomatic practices is related to a politics of connectivity, in a way that gives rise toward a sense of responsibility beyond individuals. Specifically, this is toward what Massey (1995, 2004) understands as a "geography of responsibility," i.e., as *a wider relationality where we encounter the complexity – i.e., difference – of engaging the Other in pluralized ever-changing struggles.* DeLanda (2016) following Deleuze and Guattari (1987) describes the stringy/clewy-like nature of rhizomatic substance. Further, DeLanda interprets such a connectingness as *assemblage*: an interweaving of horizontal/heterarchical co-responsibility as a "meshwork" (cf. Phillips, 2006; van Wezemael, 2008, 170–171).

- Such relational responsibility in a politics of connectivity is rooted not in one particular group or tribe, but in the emergent society of transnational networks.
- Such a solidarity transcends group or tribe. It is *a constituting reciprocal solidarity.*
- Such a sense of solidarity that is attentive to difference and complexity in pluralized networks is sometimes referred to as "*fluidarity*" (Dianne Nelson, 1999), attentive to "bleeding wounds" of precariousness and abjection in the governmentality of a heedless neoliberal global governance.

While SCMOs are seen as a kind of self-employment project and a spur to Ostrom-like circulation of *reciproquteurs* dealing with common pool resources and communing ventures, they also need to be understood in terms of social entrepreneurship financialization activities. Specifically, here they turn to a different form of networking known as *crowdfunding.* Such network financing can be understood in four forms (Moreno and Sempere, 2015):

- *equity crowdfunding* where shares/stakes in a social enterprise are purchased in exchange for perceived future economic return;
- *crowdlending* where loans are made to the promoters of a social enterprise in return for periodic income to recover initial investment, with or without interest payments;
- *reward-based crowdfunding* where funding is procured for some innovative tangible assets – cultural or technological – rather than some specified monetary return; and
- *donation-based crowdfunding* where funding is of a philanthropic character for a "cause" without any expectation of some tangible asset, such as Galicia Go Galego where the social initiative focuses on setting up and managing pre-school education.

Moreno and Sempere show a preponderance of reward-based crowdfunding for social enterprises seeking to exploit and cultivate land for goods like fruits and vegetables, wine, saffron, organic cereals and flour as well as for cosmetics, saving newspapers/magazines in digital editions.

## Tides of social insertion in Spain as a new form of social pact-ing

In the wake of the 2008 economic crisis, Spain has witnessed new spaces of social insertion – new geographies of re-embedded social responsibility. These are new spaces of self-spreading flows, spaces of a new beginning. These are new forms of social pact-ing: what we have referred to as MSSPs of social insertion involving codes of mutual referents. These pacts involve as well a re-embedding of an ensemble of new interpretive frames. Trust networking has become embedded both alongside and within the State – as bonds, as commitments, as norms, and as realized new capabilities. We turn to the post-2008 tides of social insertion known as the Indignados 15M movement as a case study of a new resilient social pact-ing, characteristic of SCMOs.

As a case study, we can turn to emergent *tides of social insertion* characteristic of new forms of solidarity-based economy, a new resilient social pact-ing in Spain – an intersubjective developing of resilience through "*commoning*," that is through pooling goods and services, and through pooling resources and social learning. They are social pacts in the *municpalismo / mutualité* tradition of Spanish syndicalism, rather than the pacts of social partnership/concerted action of neocorporatism and Ordoliberalism.

The term Tides/Mareas comes from the 15M Indignados movement that came to occupy the public squares of Spain's largest cities. Specifically, these occupations of the squares were the Tides of Social Change demonstrations in Spain following the 2008 Economic Crisis on matters other than wages. The abbreviated name "15M" refers to the date. 15M as in the 15th of May 2011, when the central squares of the major cities were occupied by tens of thousands in the start of an anti-austerity campaign bubbling up from the 50% youth unemployment (see Weiner and López, 2017). They were brought together by Marea Ciudadana (Citizens' Tides) as a coalition of 350 SCMOs. This occupation of public space was organized in citizens' assemblies and platforms.

In the following chapter, Vavarousis, Asara and Akbulut refer to this collective action as "the movement of the squares." They go on to describe how these encampments created "transitional (liminal) commons which transfigure themselves in form and substance and "spread into the urban fabric in the form of new commoning (practices and) projects after the end of the visible mobilization".

The emergent Spanish SCMOs recall Spanish traditions of municipal mutual aid initiatives (1840s–1936) within a confluence strategy of municipal, inter-urban and inter-regional social pact-ing practices. Beyond the older mutual aid practices of *mutualismo* in Spain, the new Spanish SCMOs experiment in: (1) embedding political consumerist behavior in understanding how the act of consuming is a fundamental part of the production process; as well as in (2) constituting eco-systems of voluntary stakeholders evolving toward alternative modes of sustainable living. Such *commoning* aims at fostering a cooperative process of social learning and participative decision-making. In so doing, these new Spanish SCMOs cultivate a moral economy re-embedding economic activities in durable social bonds and relations of *shared stewardship.*

In the wake of the economic crisis of 2008, Spain has witnessed what Henri Lefebvre (1974) – like Vavarousis, Asara and Akbulut in the next chapter – would call an "eruption" of liminal spaces of possibility: new spaces of self-spreading flows of new elective affinities, new spaces of becomingness and critically new spaces of social insertion. Social insertion is understood as a new form and as a new space. This is a rhizomatic moving trajectory of cascading emergent connection and reconnection.

- It is understood as *an embodied form of knowledge, and as new mediated form of politics*: an ensemble of new interpretive frames embedded and lodged both alongside and within the new Post-Franco Transition State. The Indignados movement is constituted by its own asserted and inserted social frameworks of knowledge and their signifying meaning.
- It is also understood as *a new social space of possibility*: a space of flows, a new moral economy, a new political ecology of social praxis. This is an imaginatively created symbolic space for posing and trying alternative forms of life, new participation codes and most-importantly new trust networks. Such trust-producing resources provide the "glue," holding together the Indignados as a movement for social insertion: resources that are regenerated as bonds, as shared values and norms, and as realized new capabilities.

Cristina Flesher-Fominaya (2015) details the tides of upsurging disrupting and creating of rhizomatic practices in Madrid, Barcelona, Galicia, Valencia and Aragón – making possible new types of credit pools and *mutualité*/cooperative enterprises. ***Digital interface*** between laptops, smartphones and their programs became a new point of conjuncture, and an expansion of spatial capability and autonomy in the development and dispersion of prefigurative movement conceptions, designs, rallies and occupations. This amounts to spatial reorganization – in varieties of "parallel spaces": the development of new alternatives in infrastructures of political communication, and the production of social learning. Network nodes here are not centrally coordinated; indeed they were becoming increasingly decentered/decentralized.

The concept of social insertion characterizes the movement for an increasingly shared solidarity-based economy. The intention is to rebuild new social relationships around some radical revision of the market and practices of an excessively commodified environment. Critical movement space constituted in the new social pact-ing can *scale up* the sharing of local initiatives based on what Nobel economic laureate Elinor Ostrom (1990) labels as "reciprocators" (*reciproqueteurs*) rather than as individuated entrepreneurs.

As a movement for a socially pacted democracy in Spain, the 15 M Indignados is a broad signifier foreshadowing a more confederal frame of thinking which can imaginatively anticipate and project a more flexible institutional architecture for a poly-centric, and possibly a pluri-national Spain. They harken a socially pacted New Transition, beyond the Transition of the 1978 Moncloa Pacts to a Post-Franco constitutional democracy. Such social pact-ing emerges beyond the older mutual aid of *municpalismo*. These are bootstrapping initiatives involving what Ostrom refers to as pooling of common resources and information. They involve social learning by mutual learning, and the making of multi-stakeholder pacts, i.e., codes of mutual reference.

The Indignados movements weave together a meshwork of rhizomatic practices that they come to call The Confluence of Tides. The Confluence meshwork argues for and establishes a solidarity-based economy from the bottom-up, in a way not set down by State institutions. The new social pact-ing in Spain recursively re-embeds and re-inserts the social in the sense of developing roles for sharing responsibility in adapting to managing risk in an epoch of increasing vulnerability. Rather than State-initiated concertation, the new social pact-ing from below is purely civil society initiated within their own general assemblies. The social pacts involve pooling common efforts and information (*en Común*) for bootstrapping across multiple scales, mutually setting standards and triggering sanctions. This amounts to a form of heterarchical and horizontal cooperative risk co-regulation.

The Confluence of Tides reveals a new imaginary – a genealogy of moral re-valuation in the reciprocal solidarity of more social solidarity economy. Mayor Ada Colau's Barcelona en Comú municipal government is an Indignados slate elected in May 2015. It has crystallized as a meshwork of new *commoning* SCMO projects that bridge social capital and solidary economy. Ten percent of that city's economy is based on cooperatives in 1,300 specific SCMO ventures. For example, networks like Gulf.net SemiEnergiaCoop are coordinated by Decim. Barcelona (Decide Barcelona) and Barcola web platforms for public deliberation and decision-making. In Zaragoza, La Magdalena is a solidarity economy district that brings together some 688 active cooperatives/SCMOs in the province of Aragón. By the end of 2014, those 688 represented a doubling in size in one year. The Luis Bunuel Community Center serves as an important hub, and there is strong assistance from the Zaragoza City Council and the SUSY (Social and Solidarity Economy) Project, a platform of the European Commission of the EU that links 23 national SCMO efforts from Portugal to Estonia.

Ada Colau (Colau and Alemany, 2012) declares that it is time for moving from "occupy to planning democracy." Democracy is understood not just as workplace industrial democracy, but in terms of a transformation in the social relations sustaining the capitalist mode of production. The Indignados will have to move from unmediated politics of rage and affinity groups networking to institutionalizing the new politics of Confluence. This would mean renewing the politics of the Post-Franco Transition State from a perspective of more radical democracy and more solidarity economy projects constituted from the bottom-up and framed in a synergistic network of plural visions of society.

### *Zaragoza: La Magdalena social initiatives*

In Zaragoza, La Magdalena is a solidarity economy district that brings together some 688 active cooperatives and SCMOs in the province of Aragón. By the end of 2014, those 688 represented a doubling in size in one year. The Luis Bunuel Community Center serves as an important hub, and there is strong assistance from the Zaragoza City Council and the SUSY (Social and Solidarity Economy) Project, a platform of the European Commission of the EU that links 23 national SCMO efforts from Portugal to Estonia.

Within a short space, the neighborhood of La Magdalena is significant for hosting numerous and diverse small businesses, associations and cooperatives ethically oriented both on social, democratic and environmental concerns – coffee-shops and food-shops, cooperative bookstores and book exchanges, fair trading stores. These exhibit responsible alternative consumption patterns like fair trade ecological products and food, efficient urban transit, leisure and cultural activities targeting higher social concern, empowerment of women though cooperatives developed by women.

Spain's associative networks and social movements have traditionally focused on this neighborhood. Mostly during this decade, the associative movement in La Magdelena is transforming into cooperatives. Cooperativism at La Magdalena works together with solidarity networks both:

1 at the regional and national level in organizations such as the *Red de Economía Social y Solidaria* (REAS) – Network of Social and Solidarity Economy. REAS is also called "network of networks"; and
2 in international platforms like the "Intercontinental network for the promotion of solidarity-based social economy" – *Red intercontinental de promoción de la economía social solidaria* (RIPESS).

Some outstanding initiatives is the involvement of the surrounding regional "Aragón's Social Market" (*Mercado Social de Aragón*) – MESCoop Aragón. It is part of the "Social Market" network at the national level and was one of the first solidarity markets in Spain – that includes about 44 organizations and achieves over 11 million Euros income a year. This market represents a socio-economic space shared by consumers, suppliers and distributors of products and services based on social commitment consumption (socially useful, ecologically sustainable, and being produced considering equity and democratic criteria) that also seeks to inform, train and concern society and more generally support people for example through cooperatives on financial services, work pools or social support services. Likewise, each member (social company or individual consumer) gives priority to the network when consuming and targets to close the productive cycle – also called the "circular economy."

The phenomenon is well-reflected by data: in just a year cooperatives in the metropolitan area of Zaragoza have doubled (about 500 cooperatives in the year 2014), and sum-up near 700 in the surrounding region of Aragón. Also, the map of the Solidarity Economy in Spain just below reflects the extension of this phenomenon (Figure 3.1).

This process of fast development of associations and cooperatives networks in Spain has also been reflected in the European Union

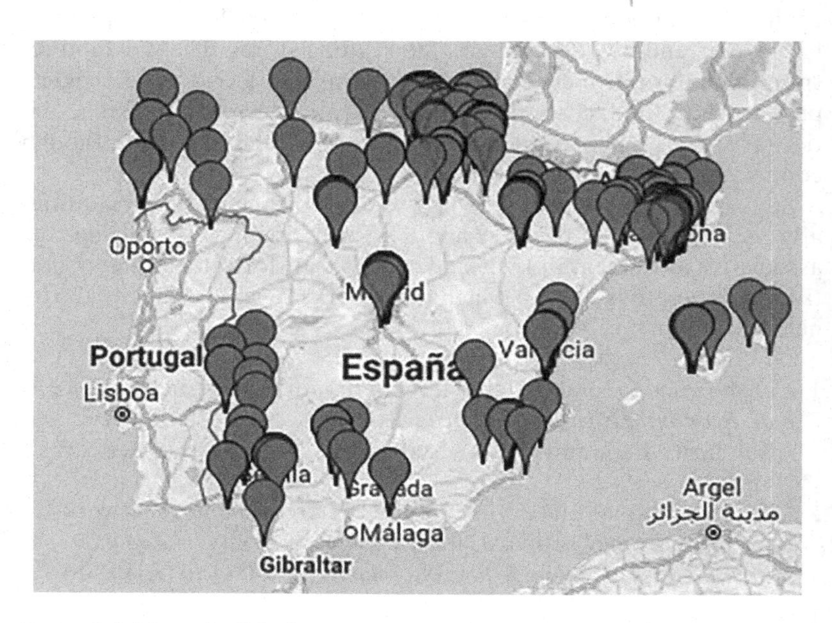


*Figure 3.1* Map of solidarity economy in Spain.
Source: www.economiasolidaria.org/entidades/mapa.

project "Sustainable and Solidarity Economy" (SUSY), depicted just below. SUSY was a three-year collaboration project between 2015 and 2018. It involved a network of 26 associations in 23 countries that aims at enhancing the competences of the new municipalism discussed below and the weaving together of transnational interurban meshwork – ten partners from Latin America, Asia and Africa. The social solidarity economy movement represents an alternative to both capitalism per se and other authoritarian State dominated economic systems. The solidarity economy movement and meshwork look to emergent best practices at the local and interurban levels. There is a focus on both advocacy and shared knowledge in seminars. The project was coordinated by European organizations involved in diverse solidarity economy experiments regarding sustainable consumption, sustainable production, solidarity food and agricultural cooperative networks, as well as micro-financing. Solidarity economy initiatives were mapped and connectingness among them was facilitated and enabled. Two million solidarity economies were estimated to exist in the European Union. At the close of the project, there was in 2018 a policy paper presentation to the European parliament, titled *Maximizing Dignity through the Social and Solidarity Economy* (Figure 3.2).

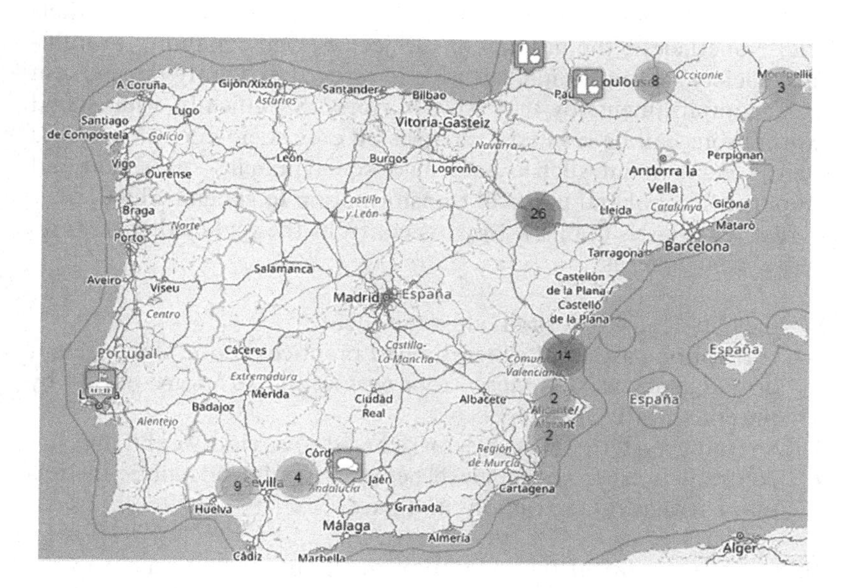


*Figure 3.2* Map of SUSY initiatives in Spain.
Source: http://es.solidarityeconomy.eu/cerca-de-ti/susy-map/.

### *The emergence of a new municipalism*

A new municipalism emerges in the collaboratitions of The New Mayors of Change, i.e., the "New Mayors"–elected in the municipal and regional elections of May 2015 in the most important Spanish cities.

The *Tides* amounted to a co-production and *insertion of the social* as a mutually referential network (meshwork) of mutual assistance horizontally constituted. They further transversally negotiated across the scales of governance, resisting neoliberal subjectifications. In the sections that follow, we detail how they prefigure a weaving together of Spanish SCMOs into *interurban meshwork …* beyond the focus on municipalismo/the new municipalism, as a "Confluence of Tides."

#### *From occupation to social insertion in public management*

As of 2015, institutional control at the local level was achieved by left-wing coalitions. This showed how the Indignados is not a spontaneous and temporary expression of citizens' mistrust and disaffection with politics, political parties and public institutions. It is rather a long-run movement with effects transcending streets demonstrations. In this new political and institutional setting, the issue is activist management capability – rather than street protests sustained by the populist rhetoric. Nonetheless, the movement's trajectory and capability to influence society's institutional setting and structures highly depends on the accomplishments by the "New Mayors." Specifically, what social expectations can be created – changes on core socio-economic issues such as employment, such as housing or social policies – need time and deep institutional and political transformation to get embedded.

Two years later, the New Mayors confront the real challenges – no matter who governs:

1 the City Halls' inherited budgetary debt;
2 corruption cases and generally bad practices (poor management or budgetary wastefulness) from past administrations (mostly by the conservative parties);
3 on-going austerity measures to reduced sovereignty debt; and
4 contradictions from the clash between ideals and public institutions' structural restrictions and pace.

Next, due to political fragmentation in the political arena, another relevant hurdle is the need of agreements with other political parties in order for policies and measures to be approved. *This is the pacting period of popular alliance politics.* This is a source of conflicts and tensions with other left-wing political parties such as the socialists or nationalists in Catalonia.

Finally, there is the relative lack of management, and the government experience of new teams heading the municipal administrations also has relevance. They could count on the support from part of the civil servants – particularly in large city halls. But it should be noted that there are also some with activist experience who do have high organizational capability to master the management of resources and working teams.

### *Safeguarding the social and the commons: the pillars of the new municipalism*

One of the main consequences of the change in municipal governments has been the fight against corruption. There is the policy of "open drawers" giving civil society (and particularly justice and judges) access to dossiers from the previous administration. Alongside, there is also the labor done by specialized police units, judges and public prosecutors which has triggered a process of dismantling criminal organizations not known until recently. These criminal syndicates have been active during the last two decades in the conservative Popular Party (and the conservative Catalan secessionist party). The magnitude reached by the corruption cases – mostly located in the regions of Madrid, Valencia and Catalonia, but reaching the national parties – raises questions about recent Spanish democratization in the post-Franco democratic period – and even the premises of history and the Spanish transition to democracy as had been told and taught.

But mostly, what is discussed – in the context of economic crisis and unpopular austerity measures – is the significance of social discourse and the need for safeguarding "the commons," the public sphere institutions and public assets. What is advocated is replacing the tight ties between politics and the business – replacing them with citizens' oriented measures, social policies and solidarity networks that go further than the municipal scale in a multi-scalar strategy. There is a push to shift from an exclusive to an inclusive public administration/management, mainly by the most vulnerable social groups.

The redefinition of institutional practices is reflected by *re-municipalization* – bringing under the local administration's control the public services that were privatized or externalized during the previous legislatures. These privatizations and externalizations are seen linked to most of corruptions cases – e.g., public transport, street cleaning – as well as in the development of new cooperative structures for new public services (such as energy supply in the Catalonia case of Som Energia).

It is worth noting the shift in the relationship between public administration and private companies – particularly big business from the financial, construction and energy sectors. There is a turn to setting as a priority "the social" and environmental concerns in the criteria for public contracting, e.g., social requirements, such as updated labor conditions of employees and updates on tax-paying.

In a sense, since the start of the economic crisis in the year 2008 "the social" has displaced "the environmental" in the public and political agenda, having an effect also in the decline of environmentalism and environmental claims. (This is the continued lack of complimentary conjoining of issues of public goods and common pool resources, that Ostrom demonstrates.) Paradoxically, in the period of the "New Mayors," institutional concern and drastic measures on core urban environmental issues – e.g., dense traffic in the city and air pollution – have reached its highest intensity. Through participation within the confluence of coalitions of social Indignados and environmentalist activists, access was gained – in most cases for the first time – to leadership of municipal governments. Many social and environmental priorities are tightly linked – such as the energy production and alternative consumption models.

## Municipalismo *as the relational open source city*

The model of the "relational city" in the geography of enhanced responsible commons-based network cooperation and deliberative democracy emerges as a result of the merging of digital peer-to-peer file-sharing networks and territorial places. This involves sharing codes for digital platforms, and thus breaking with the multinational tech company's proprietary logic of "the smart city." The result is the interurban network of "rebel cities for the common good" – to use the term of David Harvey (2012). A *geopolitics of the commons* emerges from such code connectivity.

The ascent of Mayor Manuela Carmena's Ahora Madrid coalition (2015–2019) – to the control of city hall – opened the way for a horizontal deliberative democracy network rooted in the post-2008 15M SCMOs. Carmena's regime – pushing a complementary relationship between municipal autonomy and citizen social self-management – introduced *participatory management practices* to enable the many autonomous neighborhood initiatives in Madrid. Relational open source free software and code-sharing initiatives became manifest in the rebel cities facilitating the rise of the New Mayors.

The autonomous initiative DCDCity-Aire Madrid put into effect "The Data Citizen Driven City Project" designed by the Madrid MediaLab Prado. By-passed were the sensors and a closed automated data management system installed by centralized tech multinationals. In their place, patent-free circuit boards were distributed and connected to cell phones with Android operating systems. What resulted was a community of citizen data-gatherers who get to deliberate over the issue of air quality in Madrid. The project looks to Madrid's city hall to re-initiate funding for the project.

Then there is the El Campo de Cabada self-managed space in Madrid, where a 5,500-square meter plot of land had been designated for a sports complex. The Crisis of 2008 resulted in the still-birth of the project, and the land was left empty. With the initiative of young architect collectives supported by 15M neighborhood assemblies (*circulos*), neighborhood projects took off first in online space and then in territorial place. This amounts to new local furniture enterprises, sustainable permaculture ecosystem gardening and agriculture, and community-organized cultural activities – gauging what level in multi-scalarity of markets and political arenas a cooperative should play.

### *Inter-urban networking*

Beyond social insertion, a new mode of societal differentiation and coding emerges. This emerges as an ensemble – an assemblage – of inter-urban spatially embedded node-to-node meshwork of platforms, practices and procedures. This is what the economic geographer Doreen Massey (1995, 2004) refers to as new spatial divisions of labor.

Beyond the imaginary of municipalismo, a new scalar differentiation comes to the fore: a weaving together of the bubbling up social insertion – beyond corporatist/clientelist frames. This is a weaving together of the primordial horizontal integrities, "committed to" procedures and practices of reciprocal solidarity amidst the rhizomatic practices of assemblage.

Involved here are new scalar geographies of struggle, new multi-scalar differentiations and epistemologies countering State ordoglobal interconnectivity – as the reshaping of nodal looping of metropolitan islands in an archipelago of global cities. The deeper social economic issue is the extent to which these collective practices can become enduring. That is, enduring and crystallized on a new scalar dimension of interurban meshwork. This involves a committedness to new stabilization rules for coordinating in a somewhat coherent way the negotiated network contracting and pluralist subconstitutions. This is a new form of associational life amidst imbrications within an operationally interwoven meshwork of negotiable complementary logics.

In 1990, the International Council for Local Government Initiatives (ICLEI) was formed by 200 local governments from 48 countries, having demonstrated binding connections to transnational nodes in both global supply and global distribution chains. The name was changed to ICLEI-Local Governments for Sustainability, which spawned the Urban $CO_2$ Reduction Plan, the Resilient Communities and Cities Initiative, the Local Government Climate Roadmap, and the Cities for Climate Protection (CCP) Campaign.

At around the same time in 2002, the Network for Regional Governments for Sustainable Development (nrg4SD) was represented by 50 subnational governments from 30 countries as well as 500 associations of regional governance – recognizing the joint efforts in biodiversity sustainability beyond the municipal level with regard to both conservation and development in confronting challenges like Green House Gases (GHGs) Emissions, Sewage and Safe Drinking Water. The CCP Campaign and the nrg4SD have observer status before the United Nations Framework Convention on Climate Change. The State of São Paolo has joined nrg4SD and ICLEI and embedded itself in Transnational Advocacy Networks (TANs) to influence other Brazilian States in both subnational as well as national ecological initiatives.

On $5^{th}$ of September 2015 in Spain – following the Popular Unity Confluence victories in municipal and regional elections, *Ciudades por el Buen Común* (Cities for the Common Good) laid down the foundation for the creation of an interurban meshwork in order to better coordinate "the municipalismo for change." The goal is more than municipalismo. The goal is coordination of pluralist social pact-ing initiatives regarding Social Solidarity Economy Initiatives as well as the strengthening of responsible civic participation. Further, in 2017, a Summit of Fearless Cities was held in Barcelona to pull together – rather than consolidate – these new forms of municipalismo and inter-urban associational life.

## Conclusion: beyond committedness, the ultimate question of resources

Spain has considerable solidarity-based economic initiatives. Among these initiatives SCMOs are working to construct Alternative Exchange Networks (AENs) and/or Solidarity Economy Districts (SEDs) as meshwork of economic relationships based on diffused reciprocal solidarity expressed in material practices: doing and making new products and new systems, and with it new flows and value chains of these products.

There are cooperatives from the local level to the huge Som Energia (We are Energy) – a Catalonian renewable energy consumers' coop – and to the mammoth Mondragon founded in the Basque country in 1956. Mondragon was inspired by a republican priest Don José Maria Arizmendiarrela who barely escaped the firing squad, and went on to organize a workers' cooperative making domestic heaters. The workers' cooperative grew and grew, and, by 2005, it had even acquired the French nationalized electronics corporation Brandt, with 2,000 French employees.[2]

There are also mutual societies (*mutualités*), workmen's circles, fishermen's guilds, credit unions, urban gardens, time banks, social centers for pensioners, "friendly" free shops, "popular kitchens"/food banks; occupied cinemas, alternative media, alternative social currencies, book exchanges and magical book coop called Traficantes de Sueṉos (Traffickers in Dreams) as a model of sustainable production and free downloading.

They represent four tidal waves of cooperative social insertion – also known as social self-management/*autogestion* – in Spanish history: movements of *municipalismo* and *mutualités* in the syndicalism of the middle 19th century, the Republican zones of the Spanish Civil War, the decade of the Post-Franco Transition and the 15M Indignados Movement.

There are budgetary limits to the New Municipalismo. The new post-2008 SCMOs must look elsewhere *as well* to procure necessary financial resources to keep these projects/enterprises going. All of these cooperative endeavors are represented by CEPES (Confederación Empresarial Española de la Economia Social), which lists 43,000 social enterprises in their network. There is also a Federation of the Commons for SCMOs to help in the exchange of services as well as to mutually secure financing and other resources in the manner of "club goods." Moreno and Sempere here present the case of CASX Cooperativa de l'Autofinancament Xarxa in Catalonia.

Further instances:

- These social enterprises/SCMOs are bolstered by the Law of Spanish Social Economy 5/2011 (March 29, 2011; amended in 2015) to facilitate their self-organization and enable their development.
- The acquisition and setting up of some workers' self-managed firms was supported by the unemployment benefits and severance pay of failed firms' workers who had been laid off.
- In the field of audiovisual and media social enterprises, further capital can be acquired by syndicating or selling tangible assets. (Here, the case of another Catalonian network, the CIC Cooperativa Integral which works with the Garbiana Pasessa network of three cooperatives: La Marea, MasPublico and Audiovisual Claraboia.)

The post-2008 SCMOs in Spain manifest enduring new geographies of co-responsibility. Together they function as an *assemblage* of facilitating norms of connectingness. This amounts to a rather stable operationally autonomous meshwork of mutually interdependent and mutually reinforcing practices. These prefigure and configure a new horizontalist sense of bindingness and commoning, an institutionalizing that the likes of Elinor Ostrom would understand as a reciprocal solidarity, rather than the Durkheimian vertically integrated organic solidarity.

## Notes

1 Author's reworked translation (italics added) of Karl Marx (1850), "Unser System, Oder die Weltweisheit und Weltbewegung Unserer Zeit"/Marx-Engels-Gesamtesgaube / Marx-Engels Completed Works, (MEGA-1), Berlin: Karl Dietz Verlag, 1958, p. 599.

2 Gar Alperovitz and Ronald Hanna, "Mondragón and the System Problem," Truthout, editorial, November, 2013. Alperovitz and Hannah discuss the bankruptcy of Fagor Electrodómesticos, part of the large Basque cooperative Mondragón. The bankruptcy is understood as a result of competing in the global market with Chinese companies. They stress "moving to scale": what we discuss as gauging what level in multi-scalarity of markets and political arenas a cooperative should play.

## Bibliography


Bäckstrand, Karin. 2006. "Multi-stakeholder Partnerships for Sustainable Development: Rethinking Legitimacy, Accountability and Effectiveness." *European Environment*, 16 (5): 290–306.

Castells, Manuel. 2012. *Networks of Outrage and Hope: Social Movements in the Internet Age*. Cambridge: Polity Press.

Colau, Ada and Adria Alemany. 2012. *Vidas Hipotecadas: De la Burja Immobilieария al Derecho a la. Viviendi*. Barcelona: Angle. Cf. Xavier Fina, *Els Cent Primer Dies d'Ada Colau*. Barcelona: Portic.

DeLanda, Manuel. 2016. *Assemblage Theory*. Edinburgh: Edinburgh University Press.

Deleuze, Gilles and Felix Guattari. 1987. *A Thousand Plateaus*. Trans. Brian Mussumi. Minneapolis: University of Minnesota Press.

Diani, Mario. 2003. "Simmel to Rokkan and Beyond: Towards a Network Theory of New Social Movements." *European Journal of Social Theory*, 3 (4): 387–406.

Diani, Mario and Doug McAdam. 2003. *Social Movements and Networks: Relational Approaches to Collective Action*. Oxford: Oxford University Press.

Flesher Fominaya, Cristina. 2015: "Debunking Spontaneity: Spain's 15-M Indignados as Autonomous Movement." *Social Movement Studies*, 14 (2) (2015): 142–163.; and Redefining the Crisis Redefining Democracy: Mobilizing for the Right to Housing in Spain's PAH Movement." *South European Politics and Society*, 12 (4) (2015): 465–485. Cf.

Forno, Francesca and Luigi Ceccarini. 2006. "From the Street to the Shops: The Rise of New Forms of Political Action in Italy." *South European Society and Politics*, 11 (2): 197–222.

Forno, Francesca and Paolo Graziano. 2014. "Sustainable Community Movement Organizations." *Journal of Consumer Culture*, 4 (2): 139–157.

Forno, Francesca and Paolo Graziano. 2016. "Between Resilience and Resistance: SCMOs in Italy in Times of Crisis." *European Consortium of Political Research* (ECPR), Pisa, p. 20.

Gavroche, Julius. 2014. "Anti-Capitalist Economies in Post-15M Spain." *Autonomias*. www. autonomias.org/pt/2014/08/anti-kapitalist economies in post 15M Spain-mondragon-cooperatives-okupaciones.

Gorz, André. 1977/1980. *Ecology as Politics*. Trans. Patsy Vigderman and Jonathan Cloud, Boston: South End Press. Originally published in Paris: Galilée, 1977.

Harvey, David. 2012. *Rebel Cities: From the Right to the City to the Urban Revolution*. London: Verso.

Hilson, Mary. 2011. "A Consumers International? The International Cooperative Alliance and Cooperative Internationalism, 1918–1939: A Nordic Perspective." *International Review of Social History*, 14 (2): 203–233.

Hirschman, Albert. 1970. *Exit, Voice and Loyalty*. Cambridge: Harvard University Press.

Irene Martín. "Podemos y otros.modelos de partido-movimientos." *Revista española de Sociologica*, 24 (2015): 107–114.

Lefebvre, Henri. 1974. *The Production of Space*. Trans. Donald Nicolson-Smith. Oxford: Blackwell.

Lotringer, Sylvere and Christian Marazzi. 1980. "The Return of Politics," in Sylvere Lotringer, ed. *Autonomia: Post-Political Politics / Semiotexte*, special edition, New York: Columbia University, pp. 8–20.

MacDonald, Kevin. 2002. "From Solidarity to Fluidarity: Social Movements beyond 'Collective Identity' –The Case of Globalization Conflicts." *Social Movement Studies*, 1 (2): 109–128.

Massey, Doreen. 1995. *Social Divisions of Labor: Social Structures and the Geography of Production*. Basingstoke: Macmillan.

Massey, Doreen. 2004. "Geographies of Responsibility." *Geograpfiska Annaler* 8B: 5–18.

Miller, Ethan. 2006. "Other Economies Are Possible!: Building a Solidarity Economy." *Dollars and Sense*, July/August 2006. See also www.geo.coop/node/35.

Miller, Ethan. 2013. "Community Economy: Ontology, Ethics and Politics for a Radically Democratic Economic Organizing." *Rethinking Marxism*, 25 (4): 518–533.

Monticelli, Lara. 2018. "Embodying Alternatives to Capitalism in the 21st Century." *Communication, Capitalism and Critique*, 16 (20): 501–517.

Moreno, Antonio and Salvador Perez Sempere. 2015. "Alternative Funding to Undertake Cooperatives: Case Study Spain." Paper presented at the Conference in Paris of the International Cooperative Alliance, May 27–30, 2015. Academia.edu/ 22282885.

Muino, Emilio Santiago. 2015. "Spain in Transition?: Answers from the Grassroots Facing a Collapsing Country." www.resilience.org/stories/2015-07-01/spain-in transition/.

Nancy, Jean-Luc. 1996. *Etre sigulier plural*. Paris: Galilée. Cf. Simon Critchley," With Being-With: Notes on Jean-Luc Nancy's Rewriting of *Being and Time*," *Studies in Practical Philosophy*, 1 (1), 1999: 53–67.

Nelson, Diane. N. 1999. *A Finger in the Wound: Body Politics in Quincintennial Guatamala*. Berkeley: University of California Press.

Nus, Eduard. 2016. "Strategies, Critique and Autonomous Spaces/15M from an Autonomous Perspective." 30.08.2016. www.degrowth/de/em/dim/degrowth-in-movemrnts/15M/.

Ostrom, Elinor. 1990. *Governing the Commons: The Evolution of Institutions of Collective Action*. Cambridge: Cambridge University Press.

Phillips, John. 2006. "Agencement/Assemblage." *Theory, Culture and Society*, 23 (2–3): 108–109.

Pussy, Andre. 2010. "Social Centres and the New Cooperativism of the Commons." *Affinities*, 4 (1): 176–198.

Ricouer, Paul. 2000. "The Problem of the Foundation of Moral Philosophy," in H.J. Opdebeeck, ed. *The Foundation and Application of Moral Philosophy: Ricouer's Ethical Order*, Leuven: Peeters, pp. 11–30.

Stoole, Dietland, Michele Micheletti and Jean-Francois Crépault. 2009. "The Magical Power of Consumers? Effectiveness and Limits of Political Consumerism." Paper presented for Ecpr: European Consortium of Political Research: Joint Sessions of Workshops, April 2009 in Lisbon.

Tilly, Charles. 2008. *Contentious Performances.* Cambridge: Cambridge University Press.

Van der Veen, Mara Yerkes and Peter Achtenberg, eds. 2012. *The Transformation of Solidarity: Changing Risks and the Future of the Welfare State.* Amsterdam: University of Amsterdam Press.

Van Wezemael Joris. 2008. "The Contribution of Assemblage Theory and Minor Politics for Democratic Network Governance." *Planning Theory,* 7 (2): 165–185.

Weiner, Richard R. and Iván López. 2017. *Los Indignados: Tides of Social Insertion in Spain.* Winchester / Washington: Zero Books.

# 4 The making of new commons in Southern Europe

## Afterlives of the movement of the squares

*Angelos Varvarousis, Viviana Asara and Bengi Akbulut*

### Encampment: experimentation and afterlife

The global economic crisis fiercely shook Southern European countries, with the financial and housing crisis swiftly transmuting into an economic and social crisis, on the one hand, and the length and strength of urban mobilizations it triggered in response, on the other. The mobilizations that emerged in countries such as Greece and Spain, widely dubbed as the movement of the squares, have rendered these countries the epicentre of experimentation with social and political alternatives. They have represented a challenge—at least temporarily—to the neoliberal status quo (Dikeç and Swyngedouw, 2016) through the spatialization of democratic politics (Hadjimichalis, 2013), both during and after the more visible period of square occupations. While much ink has been spilled about the period of square encampments, considerably less attention has been paid to the movements' evolution following the end of the more visible stages of mobilization and their subsequent decentralization. The few existing studies on the movements' post-square period (Arampatzi, 2017; Blanco and León, 2017; Simsa and Totter, 2017), on the other hand, focus mostly on the organizational features, forms of action, and prefigurative politics of the post-square projects and networks, overlooking the broader picture of these alternative ventures and how their emergence was related to the movement of the squares.

The overarching question that we address in this chapter is whether these movements had long-lasting social impacts, and, if they did, what sort of impacts they had. Our central argument is that the movement of the squares gave rise to a proliferation of alternative self-organized projects in some Southern European countries, which should be seen as the *social outcomes* of these mobilizations. In delineating the social outcomes, we focus on the squares' movements' *afterlives*, which denote the post-encampment periods to capture what has variously been called abeyance structures, latency periods, or submerged networks in the social movements literature (Melucci, 1996; Taylor, 1989).

More particularly, we delve on the movements of the squares and their afterlives in Athens and Barcelona. There are strong affinities between the two movements in temporal, symbolic, and organizational terms, as well as in the way they evolved. These movements operated as a source of learning and inspiration for each other, attracted a high share of previously non-politicized people, adopted similar decision mechanisms and structures, catalyzed big political changes in their respective countries and, most importantly for this study, had similar afterlives with the blossoming of self-organized (re)productive ventures. Our comparative analysis aims to answer the following questions: How did these self-organized ventures emerge from and relate to the square mobilizations? What is the broad landscape of their diffusion and the activities they are involved in? Do they constitute a continuation of the movements of the squares in a different form or are they a sign of their withdrawal?

We argue that an important form that the social outcomes took in the case of the movements of the squares is the commons, understood most generally as forms of collective governance of shared material and immaterial resources. In doing so, we combine the literatures on social movements and on the commons to explicate the social outcomes of the squares' movements. Even though both social movements and the commons are paradigmatic forms of collective action, the analytical lens of the commons has rarely been utilized in social movement studies. Conversely, the literature on the commons falls critically short in the treatment of the relation between commons and social movements. While the more recent (political-oriented) literature on the commons understands commoning as an integral part of contemporary movements (Dikeç and Swyngedouw, 2016; Stavrides, 2016;), it usually frames social movements as contentious mobilizations demanding changes and the commons as the social systems that bring about those changes[1] (e.g. Harvey, 2012) and thus fails to theorize the proliferation of the commons in the aftermath of periods of visible mobilization. Our analysis fills these gaps by bridging these two literatures.

This chapter builds on primary data collected by the first two authors in Athens and Barcelona respectively. For Athens, ethnographic research in the occupied Syntagma square between May and July 2011 was complemented by 18 in-depth, semi-structured interviews and two focus groups conducted between September 2013 and May 2014 with key movement participants involved in various encampment commissions and commoning projects. A face-to-face country-wide survey was also conducted between May 2016 and May 2017 with 404 respondents from 118 projects about commoning ventures, accompanied by 600 hours of participant observation in these initiatives.

The analysis of the *Indignados* in Barcelona is based on ethnographic research carried out over a period of three years (May 2011–May 2014) as a "partially participating observer" both during the square encampments, and the following movement decentralization in the neighbourhoods. Seventy-four in-depth interviews and six mini-focus groups were conducted in the same period with 93 activists involved in the square occupation and the commoning projects.

## Commons: the social outcomes of the movement of the squares

### *The social outcomes of social movements*

There is a revamped attention to the in-between periods of contentious mobilizations in the social movements literature. This issue is hardly new, as concepts such as "submerged networks" (Melucci, 1996), "abeyance structures" (Taylor, 1989), "spillover effects" (Mayer and Whittier, 1994), and "free spaces" (Polletta, 1999) have been put forward within the social movements literature to deal with the more latent processes in the aftermath of protests. While these conceptualizations are important for a broader understanding of movements beyond overt mobilizations, they also present shortcomings. Melucci, for instance, does not discuss how latency processes may acquire forms of visibility other than public mobilization or become a focus of overt political attention (Yates, 2015: 240). The abeyance literature, on the other hand, has stressed the passive nature of abeyance structures (Bagguley, 2002) only concerned with reproducing themselves in periods of "decline, failure and demobilization", rather than with "transforming the wider society" (Bagguley, 2002: 170). Abeyance literature hence sees movements' afterlives as only *instrumentally* important because waves of protest feed on them (see also Yates, 2015). While the "free space" concept has partly addressed this gap, it has generally been applied to small-scale settings that preceded or take place within movements (Polletta and Kretschmer, 2013) (rather than following them) and has not explored how the creation of different sites of (re)production can be an important form of everyday politicization.

The movement of the squares triggered a re-examination of the subject where a series of new concepts were employed in order to theorize the "heritage" of social movements. These have variably been referred to as Sustainable Community Organizations, Alternative Action Organizations, or Alternative Forms of Resilience (D'Alisa et al., 2015; Forno and Graziano, 2014; Kousis, 2017; Zamponi and Bosi, 2018), drawing attention to a wide spectrum of practices and organizations such as time banks, purchasing groups, cooperatives, etc., which are "organised actions and networks aimed at supporting different forms of consumption" and (re)production (Forno and Graziano, 2014: 10) rather than centring on predominantly contentious forms of actions. Bosi et al. (2016) tried to systematize these studies by providing a meaningful classification of the outcomes of social movements in political, biographical/personal, and cultural outcomes.

Building on these literatures, we propose an additional outcome category, the *social outcomes* of movements, which lie within the social field. We define social outcomes as an alternative *social infrastructure* covering different spheres of social (re)production, such as care, education, food, housing, and finance. The diffusion and "sedimentation" (Nelson, 2003) of new practices, structures, and organizations, experimented during the movements' mobilization across the social fabric, constitute the social outcomes of social movements. These are manifested in the social arena (rather than cultural, political-institutional, and biographical level), mostly affecting the meso- and group-level of collective action and societal organization (i.e. rather than the individual level as in biographical outcomes, or the political-institutional level as in political outcomes). They involve new schemes of production and reproduction, spillover effects, loose structures of solidarity and social ties, new labour unions, NGOs and service-providing organizations (often with hierarchical organizational structures), and free spaces. Following the movement continuity literature, these new schemes, networks, and structures can be conceived as the movements' *transmutation* and *continuation* beyond the more visible stages of protest, until they retain a collective identity and interconnected dense social ties among each other.

Rather than seeing the afterlives' structures in an instrumental way, we propose to view social outcomes as having a co-productive dimension in relation to the social movements and a creative dimension vis-à-vis the social fabric. Namely, the new social infrastructure, which the concept of social outcomes refers to, is created within, through and because of social movements, but is substantiated and dispersed in the social fabric both during but especially in the aftermath of the more visible stages of mobilization. In turn, the dispersed new social infrastructure can operate as the matrix where new stages of mobilization can emerge from. We seek to understand how social outcomes unfolded in the movement of the squares, which form they took, and what is their political potential.

### *Commons as social outcomes*

While social outcomes can encompass a wide spectrum of initiatives, we claim that one important social outcome took the form of commons in the case of the movements of the squares. Commons can most generally be understood as forms of collective governance of shared material and immaterial resources. More specifically, we define commons following De Angelis and Harvie (2014: 280) as "social systems in which resources are shared by a community of users/producers (commoners), who also define the modes of use and production, distribution and circulation of these resources through democratic and horizontal forms of governance (commoning)". What distinguishes commons from other types of social outcomes hence includes common management, horizontality, and direct participation of all members to decisions regarding the commons.

We frame the discussion of grassroots initiatives that followed the squares movement in our cases around commons for a number of reasons. First, utilizing the term commons has the advantage of linking these contemporary experiences with the long tradition of social systems of self-organization that includes the management of common resources (a collective fund, an urban garden, etc.). Second, the theory of the commons emphasizes the role of such grassroots ventures in the broader economic and (re)productive system and thus is suited to analyse alternative economic ventures typical of the post-square movements beyond mere alternative cultural spaces or free spaces. This has substantial implications for everyday life transformation. Third, as highlighted below, several studies on the movement of the squares have identified the production of new commons as one characterizing feature of the movement of the squares (Mayer, 2013: 14; Roussos, 2019). Finally, the term "commons" had become part of the actual vocabulary of the movement in both Athens and Barcelona (Asara, in press; Varvarousis, 2019).

Our conceptual framework addresses a critical gap that exists in the literatures on social movements and the commons. Commons has rarely been used in social movement studies as an analytical lens; conversely, the scholarship on the commons overlooks how the emergence and proliferation of the commons can be linked to social movements. Furthermore, collective initiatives that constitute the "heritage" of social movements are often seen as signs of withdrawal, depoliticization, and fragmentation rather than a form of continuation in social movement studies. For instance, Mayer (2000), in her work on (West) Germany between the 1970s and 1990s, argues that the evolution of urban movements into communal alternative projects later transformed into self-help groups and "rehab-squatting" initiatives, thus blunting their oppositional stance and leading to increasing fragmentation.

We hold that the commoning that takes place during movements' contentious phases can disseminate across the social fabric. We hold that the transmutation of social movements into commons-based experiments does not automatically signify their "death" or depoliticization. The alternative social infrastructure inherited by a social movement can develop its main properties on a larger scale rather than going into demobilization or a passive holding process. Obviously there is no absolute rule to judge the effects of the commons on depoliticization or movement fragmentation. Alternative (solidarity) spaces can transport political meanings and ideas through their practices and discourses (Yates, 2015), performing an empowerment role vis-à-vis austerity (Arampatzi, 2017). In a Rancierian fashion, these forms of action can be held to constitute political action whenever they involve the disruption and disidentification of the existing naturalized order, and the reconfiguration of what is acceptable and visible through the constitution of a collective political subject based on equality (Rancière, 1999). In addition, the commons-based experiments play a politicizing role on participants′ everyday life and can operate as active networks for the emergence of new mobilizations, pointing to a co-productive relation between the commons and social movements in our cases.

We substantiate these arguments through the empirical analysis of the movement of the squares and their afterlives in Athens and Barcelona in the following sections.

## *Transitional commoning spaces in Athens and Barcelona*

Common space was "secreted" (Stavrides, 2016) in both cities during the encampments. Numerous commons with similar characteristics were created alongside the new common space: social kitchens; self-organized medical centres; and libraries, art and cinema spaces, cleaning commissions, kindergartens, time banks, exchange bazaars, and even a community garden (in Barcelona) symbolized the collective effort to reinvent everyday life "in common" by offering a basic (re)productive social infrastructure. A number of content-related commissions fostering discussions on core socio-political issues were also created, such as education, unemployment, direct/real democracy, the economy, and the environment.

In both cases the General Assembly was the supreme decision-making body on both practical/organizational and content-related matters that affected the whole movement. It also operated as the central ritual of encampments. Both Syntagma Square and Plaça Catalunya became the epicentres of incipient urban commoning practices. Especially after the squares were left, popular assemblies were meeting independently in their neighbourhoods and periodically convening at the general assembly of Syntagma Square (in Athens) and the Inter-Neighbourhood Coordinating Space (INCS, in Barcelona). The discussions taking place in the general assembly/INCS fed back to the neighbourhood assemblies, though without binding them. This emerging decentralization-recentralization mechanism evolved into a network of interconnected commoning projects that were dispersed throughout the metropolitan space.

The square commons operated as transitional or liminal common spaces (Varvarousis, 2018; Varvarousis and Kallis, 2017), facilitating interactions between diverse people and groups; they were subsequently transmuted into a series of new projects that diffused in the urban fabric of Athens and Barcelona.

## *"Squaring" the city: the social outcomes in Athens and Barcelona*

### *The transmutation of the social movement in the social fabric*

The mechanism of emergence and diffusion of the new commons from the squares' transitional commons took direct and indirect forms. The "direct mechanisms" consist of three forms. First, some of the commoning projects created in the squares continued their activities in neighbourhoods after the mobilizations ended, in many cases with the same membership and name. We call this the *transplantation* process. Second, certain commoning projects were conceived during the encampment periods, but materialized only in their aftermath. We call this the *ideation* process. Third, numerous ventures sprouted as autonomous projects conceived in the bosom of neighbourhood assemblies. We call this the *breeding* process. Finally, the "event" of the squares also created a series of imaginary and discursive resources (Khanna, 2012), which gave birth to a new public culture (Pantazidou, 2013), a different "social climate" (Fernández-Savater et al., 2017) or even, as

Karyotis (2018: 30) put it, "a Plan C based on the social reorganization around the commons", operating as a generator of commons through indirect mechanisms.

In Athens, neighbourhood commoning initiatives had been marginal before 2011: the few ones created during the 2008 December Revolt were generally inactive and reactivated during the movement of the squares. Preexisting political networks, organizations, and radical parties played a role in their stronger diffusion in the aftermath of the movement of the squares. Survey findings indicate that the majority (54%) of the new commons was created by groups of individuals, although a substantial share (20%) was created directly by radical political groups/parties—such as Solidarity4all, a network dedicated to promoting commoning projects, founded and funded by Syriza, the left-wing party in power in Greece between 2015 and 2019.

In Barcelona, solidarity economy initiatives, collaborative consumption/production practices, and other civic networks that existed before the crisis (Conill et al., 2012) as social outcomes of the past alterglobalization, squatters' and neighbourhood movements (Castells, 1983; Flesher Fominaya, 2015, 2017) also played a role in the diffusion of new commoning projects. In particular, an important role was played by occupied spaces (e.g. Can Masdeu, Can Vies) and Neighbourhood Associations.

### *Social outcomes in Athens and Barcelona*

The boom of commoning projects in Athens has been connected to the Syntagma mobilizations by some scholars (Theocharis, 2015; Varvarousis and Kallis, 2017). Our survey confirms this: 51% of respondents stated that the ventures they participate in are direct outcomes of the social movements in Greece between 2008 and 2011, while only 23% considered their projects unrelated.

The new projects encompass a wide spectrum of ventures in various fields of production and reproduction such as self-managed social clinics and pharmacies, worker cooperatives, urban occupations, time banks, alternative currencies and solidarity exchange networks, urban gardens, farmer/consumer cooperatives, farmer markets without intermediaries, and art/publishing collectives and an occupied factory. This "explosion" indicates that a horizontal solidarity and commoning inherited at least partly from the square mobilizations were established as resilience systems to cope with the effects of the crisis and the dismantling of the welfare state.

Social clinics, for instance, aim to provide healthcare to those excluded from the public health system; some struggle also against austerity and attempt to develop an alternative healthcare model. These clinics hardly existed before 2011, but multiplied afterwards: of 72 known initiatives across Greece in 2014 the majority were started in 2011–2012 (Adam and Teloni, 2015). Likewise, solidarity hubs, mainly active at the neighbourhood level, are involved in the distribution of food, clothing, and free lessons. While very few existed before 2011, there were over 110 in 2014 (solidarity4all 2014). Direct producer-to-consumer networks also flourished after 2011: virtually non-existent or unknown before 2011, there were 47 recorded networks across Greece in 2014 (Solidarity4all 2014). Finally, 70% of existing social-solidarity economy organizations in the country was created after 2011, of which 36% are based in Athens (Varvarousis et al., 2017).

Barcelona similarly witnessed a surge of new commons in the aftermath of the square occupation. This included time banks, social centres, free social canteens for the disadvantaged, food banks, second-hand barter markets organized by neighbourhood assemblies, community gardens, anti-eviction assemblies and networks, and agro-ecological producer-to-consumer groups (doubling between 2011 and 2015 in Barcelona, Fernàndez and Miró, 2016: 103). Cooperatives witnessed an increase of more than 3,000 workers in the years 2011–2015 (ibid.: 14). The annual Catalunya Fair (and Network) of Solidarity Economy was created one year after the encampments. Currently there are around 4,700 social-solidarity economy initiatives (860 cooperatives and 260 community economies) in Barcelona, involving about 500,000 members and 100,000 volunteers (Fernàndez and Miró: 11).

But how the transitional commons of the squares gave birth to these new commons?

In both cities, discussions about the aftermath had already begun during the encampment period. In Athens, this happened within a context of increasing police violence in early July. The proposals varied from disseminating the squares′ political practices throughout Athens to strengthening existing neighbourhood assemblies, creation of solidarity hubs of commoning practices, and maintaining the square occupation at any cost. In Barcelona, instead, the General Assembly decided to leave Plaça Catalunya and decentralize in the neighbourhoods to focus on "the creation of alternatives", despite some conflicting views over timing (Asara, 2016).

Despite these differences, the dissemination of encampments' transitional commons within the urban fabric involved both direct and indirect mechanisms explained above. In Athens, the *transplantation* process occurred with some square projects such as the artist collective, the time bank and the exchange bazaar, which continued their activities for months (or even years) after the end of mobilizations. The first network of community gardens in Barcelona was created following the interactions while planting the community garden in the square (Asara and Kallis, 2019).

The *ideation* process, on the other hand, is embodied in the peri-urban eco-community project "Spithari Waking Life" and the "Metropolitan Social Clinic" that treats thousands of patients yearly, both of which were conceived during the occupation but materialized in its aftermath in Athens. A series of projects were conceived during the encampment period but materialized afterwards also in Barcelona, such as the CASX, a self-managed, no-interest social finance cooperative.

However, the majority of ventures were created through a *breeding* process in Barcelona. An example is Ateneu Cooperatiu La Base, an umbrella project (manifold commons) resembling a social centre, including sub-commons such as an agroecological consumption cooperative and a co-maternity nursing group. Other examples include Recreant Cruïlles, a self-managed project including a community garden and cultural spaces (Asara and Kallis, 2019), and the Gracia squatted bank, a space for meeting, discussing and exchanging second-hand clothes. The squatted Poblenou Indignant community gardens were founded for the use of both movement activists and neighbours, and were directly inspired by the square community garden. Similarly in Athens, "Solidarity of Pireaus", which experiments with various forms of commoning like social kitchen, manufacturing, and (peri)urban farming, was created by square participants who wanted "to do something similar in their neighborhood", as put by a leading figure of the organization. A social clinic member in Athens similarly stated, "Our own social clinic was, at least to some extent, a demand derived from the squares (…) it was among the objectives set by the movement participants".

Hundreds of new projects were also created with the "square effect" on "public culture" (Pantazidou, 2013), e.g. through indirect mechanisms. This was strengthened by pre-existing social infrastructure of movements. As mentioned earlier, only 23% of survey respondents consider their projects unrelated to the movements in Greece between 2008 and 2011; of those, a fraction participates in projects that existed before those movements. A 2017 study likewise showed that 70% of social-solidarity economy initiatives were launched immediately after the movements of 2011 (Varvarousis et al., 2017).

While we do not have comparable data for Barcelona, our ethnographic research indicates that the creation of new commons was indirectly boosted by the new type of social climate created by the Indignados. The most prominent case is Can Batlló, a 14-ha former factory space hosting numerous commons, including socio-cultural spaces (an auditorium, a self-managed library, etc.), a (book) publishing, a brewery and a (co-)housing cooperative, a cooperative incubator, carpentry, blacksmith, and construction spaces (Asara, 2019). While Can Batlló was part of a 40-year struggle, its self-management could only be initiated one month after the square occupation. Other noteworthy examples include Ateneu Flor de Maig, a self-managed socio-cultural space squatted in autumn 2012, historically an important consumption cooperative; the Farigola square, a common space creating an "unofficial" square; and the Xino garden, whose participants have been increasing after the square occupation.

## Conclusion: the political character of commoning

This chapter has argued that an important social outcome of social movements has been the creation of commons onto which the movements transmuted in the aftermath of the wave of protest. It has further identified, by a comparative study, the mechanisms through which commons were forged from the liminal commons of the squares and demonstrated how transitional (liminal) commons spread into the urban fabric in the form of new commoning projects after the end of the visible mobilizations. Of course, these ventures are not completely novel or different from those emerged from earlier protest cycles of the 1970s and 1980s in countries such as Germany, France, or Italy. They differ, however, in their increased focus on the economic aspect (i.e. aim to guarantee means of livelihood, employment, or the initiative's economic autonomy), their inclusiveness, and their proactive engagement with the neighbourhoods and the cities they are located in.

A further question in this context concerns the political character of these new commons and the extent to which they sustained, advanced, and spatialized the political vision of the movement of the squares. This is a compelling question especially in the light of the post-political literature which claims that "the political" (contestation and agonistic engagement) is increasingly colonized by politics (technocratic mechanisms within an unquestioned political-economic framework) (Wilson and Swyngedouw, 2014: 6) attested by the erosion of democracy and a consensual mode of governance. For some theorists, the commons we discuss here are driven by self-realization and expressive politics, and are thus instances of the post-political condition (e.g. Blühdorn, 2017).

While we take the post-political critique seriously, we hold that the projects we examine here go beyond self-realization and have rather been motivated by the need to "spatialize" (Dikeç and Swyngedouw, 2016) the movement's values and practices onto the urban fabric. In both of the cases analysed here, contribution to social and political change, creating a different politics in daily life and transforming the society were expressed as important motivations for participating in such initiatives by a majority of participants. In addition, these commons often engage proactively with the wider neighbourhood and city, through which they turn into places of resistance from which to launch strategies to counter austerity politics (see also Blanco and León, 2017). For example, Ateneu La Base in Barcelona fights against urban plans that are likely to induce further gentrification in the neighbourhood and Can Batlló collaborates with the Neighbourhood Association to get the local administration to comply with its commitment to build public services. Neighbourhood assemblies also directly collaborate with the housing movement (PAH) in contesting austerity cuts and evictions (Blanco and León, 2017). In Athens, on the other hand, the alternative social structure created by the new commons became the basis upon which the new "solidarity with refugees" movement evolved within the context of massive migrant flows of 2015 and 2016 (Varvarousis, 2018). Finally, these initiatives often brought about a (re)politicization and involvement of groups previously politically inactive and/or unmoved by national and central politics through their active and open engagement with the neighbourhood and politics as well as their insistence on keeping a local focus.

For us, the commons do not represent a "hibernation" of the movements in a passive stand-by mode but rather allow an important transformation of everyday life through the different modes of consumption, production, organization, and being-in-common they demonstrate. Commons as social outcomes are also not the mere "heritage" of mobilizations, but also often operate as bases for new mobilizations and contentious politics. They are political not in the sense of their eventual effectiveness in challenging the dominant order, but in their actual practices and capacity of politicizing the particular issues they address. They visibilize and problematize the links between the specific issues, practices, and activities they focus on with broader and structural dynamics of injustice, inequality, and exclusion.

## Note

1 For a more comprehensive literature review of the usage of social movements in the commons literature see Varvarousis (forthcoming).

## References

Arampatzi, A. (2017) The spatiality of counter-austerity politics in Athens, Greece: Emergent 'urban solidarity spaces'. *Urban Studies*, 54(9): 2155–2171.

Asara, V. (2016) The Indignados as a socio-environmental movement. Framing the crisis and democracy. *Environmental Policy and Governance*, 26(6): 527–542.

Asara, V. (2019) The redefinition and co-production of public services by social movements: The Can Batlló social innovation in Barcelona. *PArtecipazione e Conflitto*, 12(2): 539–565.

Asara, V., and Kallis, G. (2019) Prefigurative territories: The production of space by the Indignados movement. Paper under review.

Asara, V. (in press) Untangling the radical imaginaries of the Indignados' movement: Commons, autonomy and ecologism. Paper under publication in Environmental Politics.

Bagguley, P. (2002) Contemporary British feminism: A social movement in abeyance?. *Social Movement Studies*, 1(2): 169–185.

Blanco, I., and León, M. (2017) Social innovation, reciprocity and contentious politics: Facing the socio-urban crisis in Ciutat Meridiana, Barcelona. *Urban Studies*, 54(9): 2172–2188.

Blüdorn, I. (2017) Post-capitalism, post-growth, postconsumerism? Eco-political hopes beyond sustainability. *Global Discourse*, 7(1): 42–61.

Bosi, L., Giugni, M., and Uba, K. (2016) *The Consequences of Social Movements*. Cambridge: Cambridge University Press.

Castells, M. (1983). *The City and the Grassroots: A Cross-Cultural Theory of Urban Social Movements (No. 7)*. Berkley: University of California Press.

Conill, J., Castells, M., Cardenas, A., and Servon, L. (2012) Beyond the crisis: The emergence of alternative economic practices. In: Castells, M., Caraça, J., and Cardoso, G. (eds.), *Aftermath: The Cultures of Economic Crisis* (pp. 210–250). Oxford: Oxford University Press.

D'Alisa, G., Forno, F., and Maurano, S. (2015) Grassroots (economics) activism in times of crisis: Mapping the redundancy of collective actions. *Partecipazione e conflitto*, 8(2): 328–342.

De Angelis, M., and Harvie, D. (2014) The commons. In Parker, M., Cheney, G. Fournier, V., and Land, C. (eds.), *The Routledge Companion to Alternative Organizations* (pp. 280–294). Abington: Routledge.

Dikeç, M., and Swyngedouw, E. (2016) Theorizing the politicizing city. *International Journal of Urban and Regional Research*, 41(1): 1–18.

Fernàndez, A., and Miró, I. (2016) *L'economia social i solidària a Barcelona*. La Ciutat Invisible, SCCL and Comissionat d'Economia Cooperativa, Social i Solidària – Ajuntament de Barcelona. Gener, Barcelona.

Fernández-Savater, A., Flesher Fominaya, C., Carvalho, L., Çiğdem, H., Elsadda, W., El-Tamami, P., Horrillo, S., and Nanclares, Stavrides, S. (2017) Life after the squares: Reflections on the consequences of the occupy movements. *Social Movement Studies: Journal of Social, Cultural and Political Protest*, 16(1): 119–151.

Flesher Fominaya, C. (2015) Debunking spontaneity: Spain's 15- M/Indignados as autonomous movement. *Social Movement Studies*, 14(2): 142–163.

Flesher Fominaya, C. (2017) European anti-austerity and pro-democracy protests in the wake of the global financial crisis. *Social Movement Studies*, 16(1): 1–20.

Forno, F., and Graziano, P. (2014) Sustainable community movement organisations. *Journal of Consumer Culture*, 14(2): 139–157.

Hadjimichalis, C. (2013) From streets and squares to radical political emancipation? Resistance and lessons from Athens during the crisis. *Human Geography*, 6(2): 116–136.

Harvey, D. (2012) *Rebel Cities. From the Right to the City to the Urban Revolution*. London: Verso.

Karyotis, T. (2018) Beyond hope: Prospects for the commons in austerity-stricken Greece. In: Holloway, J., Nasioka, K., and Doulos, P. (eds.), *Beyond Crisis: After the Collapse of Institutional Hope in Greece, What?* (pp. 19–53). San Francisco: PM Press.

Khanna, A. (2012) Seeing citizen action through an unruly lens. *Development*, 55(2): 162–72.

Kousis, M. (2017) Alternative forms of resilience confronting hard economic times: A southern perspective. *PArtecipazione e COnflitto*, 10(1): 119–135.

Mayer, D., and Whittier, N. (1994). Social movement spillover. *Social Problems*, 41(2): 277–298.

Mayer, M. (2000) Social movements in European cities: Transitions from the 1970s to the 1990s. In: Bagnasco, A., and Le Galès, P. (eds.), *Cities in Contemporary Europe* (pp. 131–152). Cambridge: Cambridge University Press.

Mayer, M. (2013) First world urban activism. *City*, 17(1): 5–19.

Melucci, A. (1996) *Challenging Codes. Collective Action in the Information Age*. Cambridge: Cambridge University Press.

Nelson, L. (2003) Decentering the movement: Collective action, place, and the 'sedimentation' of radical political discourses. *Environment and Planning D: Society and Space*, 21: 559–581.

Pantazidou, M. (2013). Treading new ground: a changing moment for citizen action in Greece. *Development in Practice*, 23:5–6, 755–770.

Polletta, F. (1999) "Free spaces" in collective action. *Theory and Society*, 28(1): 1–38.

Polletta, F., and Kretschmer, K. (2013) Free spaces. In: Snow, D., Della Porta, D., Klandermans, B., and McAdams, D. (eds.), *Wiley-Blackwell Encyclopedia of Social and Political Movements*. Malden: Wiley-Blackwell, https://onlinelibrary.wiley.com/doi/abs/10.1002/9780470674871.wbespm094.

Rancière, J. (1999) *Disagreement: Politics and Philosophy*. Minneapolis: University of Minnesota Press.

Roussos, K. (2019) Grassroots collective action within and beyond institutional and state solutions: The (re)politicization of everyday life in crisis-ridden Greece. *Social Movement Studies*, 18(3): 265–283.

Simsa, R., and Totter, M. (2017) Social movement organizations in Spain: Being partial as the prefigurative enactment of social change. *Qualitative Research in Organizations and Management*, 12(4): 280–296.

Stavrides, S., (2016) *Commons Space*. London: Zed Books.
Taylor, V. (1989) Social movement continuity: The women' s movement in Abeyance. *American Sociological Review*, 54(5): 761–775.
Theocharis, Y. (2015). Every crisis is a digital opportunity: the Aganaktismenoi movement's use of social media and the emergence of networked solidarity in Greece. In: *The Routledge Companion to Social Media and Politics* (pp. 184–197). Routledge.
Wilson, J. and Swyngedouw, E. (2014). Seeds of dystopia: Post-politics and the return of the political. In: J. Wilson and E. Swyngedouw (eds), *The Post-Political and Its Discontents: Spaces of Depoliticisation, Spectres of Radical Politics* (pp. 1–22). Edinburgh: Edinburgh University Press.
Varvarousis, A., and Kallis, G. (2017) Commoning against the crisis. In: Castells, M. (ed.), *Another Economy Is Possible. Culture and Economy in a Time of Crisis* (pp. 128–160). Cambridge: Polity Press.
Varvarousis, A., Temple, N., Galanos, C., Tsitsirigos, G., and Bekridaki, G. (2017) *Social and Solidarity Economy Report*. Athens: British Council Publications.
Varvarousis, A. (2018) *Crisis, Commons and Liminality*. PhD Thesis.
Varvarousis, A. (2019) Crisis, commons and decolonization of the social imaginary. *Environment and Planning E: Nature and Space*, 2(3): 493–512.
Varvarousis, A. (in press) The rhizomatic expansion of commoning through social movements (Paper under publication in Ecological Economics).
Yates, L. (2015) Everyday politics, social practices and movement networks: Daily life in Barcelona's social centres. *The British Journal of Sociology*, 66(2): 236–258.
Zamponi, L., and Bosi, L. (2018) Politicizing solidarity in times of crisis: The politics of Alternative action organizations in Greece, Italy, and Spain. *American Behavioral Scientist*, 62(6): 796–815.

# 5 Shopping for a sustainable future

## The promises of collectively planned consumption

*Alice Dal Gobbo and Francesca Forno*

### Introduction

The present is riddled with environmental problems ranging from climate change to resource depletion to waste management; with a lasting economic stagnation that signals how many countries have never recovered from the 2008 financial crisis; with the upsurge of nationalistic, violent governments and declining participation in institutional politics. More and more, the valorisation processes that are at the basis of capitalist accumulation dig deep and large into life, commodifying always novel fields of its reproduction. Commodification takes extractivist forms, so that we witness a decoupling between economic growth and well-being for people, but also local and global ecologies: expanding economy means, in some ways, depleting life (Harvey, 2007; Dal Gobbo and Torre, 2019; Leonardi, 2019). The survival of this economic model means, even in the everyday life of many people in wealthy countries, social and economic insecurity, fragmentation, disengagement, lower levels of well-being and physical health. Widespread discontent makes the need and wish for social change very urgent; and yet the traditional ways of participation into political debate and emancipation seem largely closed to the voices of citizens.

The aim of this chapter is to reflect on the ways in which grassroots mobilisations try to articulate and respond to these challenges. In particular, it looks at how Sustainable Community Movement Organisations (SCMOs) emerge and grow stronger as a response to the present, multi-faceted, crisis by looking in depth at one particular form of consumer groups which has spread in Italy over the last 20 years: the Gruppi di Acquisto Solidale (Solidarity Purchasing Groups, hereafter GAS). By directly addressing basic necessities of everyday life, such groups go to the heart of many of the systemic dysfunctions and contradictions that characterise the present modes of production, distribution and consumption (see Forno and Graziano, this volume).

Through the use of data from a mixed-methods research, the chapter considers in what sense GAS groups can be defined as ways to "organise for a sustainable future" by highlighting the processes through which they constitute original "assemblages" (Deleuze and Guattari, 2014): relatively patterned material flows *and* symbolic-discursive constructions that are ongoingly built by human and non-human beings in interaction. These assemblages not only give concrete alternatives to the current unsustainable system, but indeed embody a novel style of doing politics. Our argument begins with an outline of the historical emergence of SCMOs in Italy and a brief description of what GAS are and how the research has been conducted. Our discussion engages with the transformative potential of GAS's politics, while we end with a critical assessment of the opportunities and limitations of this form of grassroots mobilisation.

## The world of SCMOs in Italy

Especially since the mid-1990s, Italy has seen the spread of many initiatives and groups promoting political consumerism and collective practices of sustainable procuring and provisioning as a way to foster social change. The increase of different forms of activism aiming to promote alternative ways of consuming, trading, and spending observed during the 1990s has often been connected to the rise of the Global Justice Movement (GJM). By identifying the market as a privileged arena for political activism, the GJM in fact opened up space for experimentation with consumerist action (Micheletti, 2003; della Porta, 2006). It is during this period that, in Italy, market-based and lifestyles actions begun to extend to an increasingly large number of people and to be incorporated among the repertoire of action of many different social movement organisations. A process that Bosi and Zamponi (2015) refer as a spillover effect (Meyer and Wittier, 1994) between the GJM and the more recent social movement landscape in Italy with the transmission of a broad set of practices, referring in particular to market-based activism.

In other words, and in line with what happened in other Western countries, after the GJM, also in Italy, market-based action began to be incorporated in the repertoire of action of a variety of groups, not only those specifically involved in the promotion of changes in consumption habits and behaviour but also by pacifist, religious, workers' and anti-mafia groups (Forno and Gunnarson, 2011; Micheletti and McFarland, 2011). Moreover, besides Fair Trade organisation, aiming to build transnational awareness across borders and to step up pressure on corporations, over the years Italy has seen the rise and spread of a number of different initiatives more often acting at local level (Forno and Graziano, 2014). Examples of such initiatives are new consumer-producer cooperatives, community-supported agriculture, solidarity purchasing groups, ecovillages, voluntary simplicity, slow food, barter groups, urban gardening, degrowth and transition town networks.

Beyond promoting and practicing solidarity market exchanges, a common trait of such initiatives appears to be their use of market-based action that goes beyond boycotting and buycotting, with a focus on the creation of alternatives, articulating economic concerns with calls for alternative ways of living (Asara et al., 2015). In other words, instead of appealing to formal (local, national, international) institutions by lobbying and/or putting pressure so as to make them change their political decisions, as we will see in the case of GAS, such organisations tend to act locally by ongoingly building concrete alternatives to the system they are contesting. In some ways we might say that instead of asking for change, they produce the change itself in the form of alternative ways of socio-ecological and economic organisation, establishing novel material and cultural-symbolic patterns. In this sense, they are "prefigurative": they try to embody an alternative world that attempts to become concrete for people outside of their "niche". Privileging direct interaction and rejecting higher-level organisation, in fact, does not imply that they cannot scale up through replication, the creation of networks and alliances.

The financial crisis and the ensuing austerity policies seem moreover to have further fuelled the growth of innovative consumerist practices (Guidi and Andretta, 2015; Forno and Graziano, 2019). In what follows, we will critically evaluate these features – their opportunities for building a significant transition to sustainability – through the example of Solidarity Purchasing Groups (GAS).

## The GAS in Italy and the research context

GAS are "groups of consumers who purchase collectively through a direct relationship with producers, according to shared ethical principles (the 'solidarity' concept)" (Brunori et al., 2012: 31). These are mainly social and environmental justice as well as the search for a price that is equitable for both consumer and producer. The first GAS was established in 1994, and over the last decades they have spread throughout Italy – even though there is a greater concentration in the northern regions of the country. GAS may collectively buy bread, pasta, flour, milk, dairies, oil, fish, meat, detergents, wine, preserves, juices and jams, fruits and vegetables and other items of everyday use (such as detergents and basic toiletries). They also purchase textiles and "alternative" services such as renewable energy and sustainable tourism experiences.

Unlike other collective purchasing groups, GAS do not simply aim for the cheapest price, but choose products and producers with the explicit goal of building a viable alternative to the "consumer society" that they regard as exploitative, unjust and unsustainable – economically, socially and environmentally. They thus search for local products in order to minimise the environmental impact of transportation and to support regional economies and to increase the transparency of the food chain. They also favour organic agriculture and reusable or eco-compatible goods for environmental reasons, fair-trade goods to help disadvantaged producers (Signori and Forno, 2019).

Being part of a GAS requires high levels of commitment and involvement since groups self-organise. Each member (*gasista*) is expected to take responsibility for one or more of the products purchased; further, they attend regular meetings where shared norms are developed, financial and logistical aspects are managed, activities are planned. At the local level, once these groups achieve a critical mass, they sometimes network together into *Distretti di economia solidale* (Districts of Solidarity Economy) that are roughly similar to the Anglo-Saxon transition town movement and have explicit governance objectives (Grasseni, 2014).

Drawing on a mixed-methods study on GAS groups and members in Lombardy, in the following pages we aim to shed further light on the political aspects of GAS participation, which we will try to conceptualise as a process of assemblage-building. Our main argument will be that, differently from the models of politics that have been hegemonic throughout the modern period, GAS constitute as a "prefigurative" practice that embodies socio-ecological change by reshaping symbolic and material flows within the concrete here-and-now of everyday experience, while simultaneously reaching beyond this local level itself to build trans-local alliances. We start with a few considerations on GAS as ways of responding to the aforementioned crisis, we then move on to the ways in which they seek (in their peculiar ways) to build "new worlds" – and how this repositions the role of "responsibility" and desire within politics for sustainability.

## GAS politics as the making of sustainable assemblages

As recent research has argued, the rise and spread of grassroots initiatives aiming at building alternative and sustainable networks of

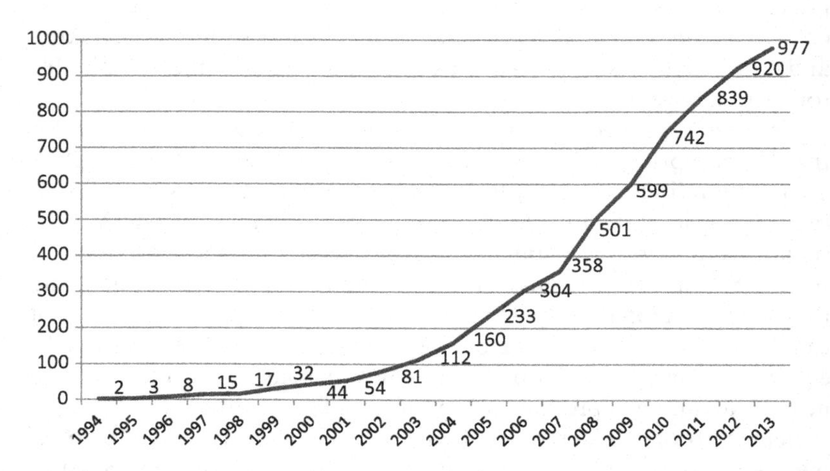


*Figure 5.1* Growth trend of Solidarity Purchase Groups in Italy, 1994–2013: from 1 to 1,000 GAS self-registered on www.retegas.org.

production, exchange and consumption are strictly linked with times of economic crisis (Castells et al., 2012; Bosi and Zamponi, 2015; D'Alisa et al., 2015; Kousis, 2017; Giugni and Grasso, 2018). In line with other SCMOs, the trend followed by GAS shows a significant increase in numbers after the inception of the 2008 financial crisis (see also: Forno and Maurano, 2016; Matacena, 2016; Forno and Graziano, 2019). We argue that this increase is not accidental but it has to do, on the one hand, with the fact that diminishing purchasing power requires people to find ways to make everyday shopping sustainable; on the other, the trend signals a shift in modalities of political participation that has been favoured by the crisis. We address the two issues in turn (Figure 5.1).

## *GAS and the experience of the crisis*

As shown in Forno, Grasseni and Signori's research on GAS (2015), *gasistas'* profile is that of the educated, but not necessarily rich, middle class (60.1% of them have white-collar jobs or are teachers). They hold rather high "cultural capital", but they are not particularly high-income earners. The financial crisis has deeply impacted this population group via unemployment, diminishing purchasing power and rising prices. Simultaneously, educated middle-class families have tended, in time, to build share narratives and practices about what "good" or proper shopping is – involving issues of fairness, equity, worker's rights and lately environmental sustainability. As the market increasingly absorbs and sells these values through commodities, at increasing prices, ethical consumption within mainstream market distribution becomes economically unsustainable for most. GAS give the opportunity to purchase ethically verified, environmentally sustainable and transparent products at a cheaper price than their equivalents in large organised distribution. Even though purchases are more expensive than discount or supermarket ones (one of the reasons why they remain a classed practice), by shortening the distribution chain, taking away intermediaries between producer and consumer, and organising purchasers they manage to reduce costs and thus make "ethical consumption" feasible even for people who are not in the upper income levels.

But beyond the issue of affordability, GAS practices are indicative of a wider response to the multiple dimensions of the present crisis – one in which economic issues are accompanied by ecological, socio-political and psychological ones. Sedimented and taken-for-granted forms of material and symbolic organisation are unsettled, while the dissatisfaction with existing, unequal and exploitative forms of life grows. The desire of "*changer la vie*" (Lefebvre, 2014) emerges. Yet, the present is partially "closed" to opportunities of innovation and transformation emerging from the grassroots level of civil society. Individualisation and decline of traditional forms of public participation undermine the opportunities to articulate and struggle for people's needs. Economically, growing inequalities mean that more and more actors are increasingly less able to have any influence on market regulation and decision-making. Politically, traditional institutions have proved unable to concretely respond to claims for environmental and social justice either on local, regional, national or international levels.

By embodying a way of practicing politics that is alternative to mainstream institutionalised ones, GAS are the sign of (and one response to) the need for new emancipations. Interestingly, indeed, *gasistas* are fairly concerned about political issues, but few of them trust the formal institutions where politics have articulated in the last decades. Their search for effectiveness in the present socio-political impasse and their desire to *concretely* participate and act for common causes are among the main drivers for joining GAS. The latter can then be regarded as alternative ways of political engagement, in which people look for new relationships and material responses to environmental problems through their acts of consumption.

### *Building a new political space: making worlds*

Participating to GAS means *building* and *making* something new – new systems of production, provisioning and socialising around everyday necessities (especially food but also detergents, household goods, clothes, etc.). Differently from asking and petitioning (which symbolically appeal to institutions for them to make a change), this practice is both material and symbolic in the sense that it seeks to embody desired changes in the here and now of concrete existence. In this, GAS are "prefigurative" (Schlosberg and Coles, 2016): they show one possible way of constructing new socio-ecological relations, new worlds; they articulate and assemble weaving together discourses and things, words and objects, ideologies and material flows. Because they do not address one or the other side of these (fictitious) dichotomies, GAS are one interesting and arguably successful way of building new economies, ecologies and socialities.

A case in point is the way in which GAS's collegial meetings take place, and their function. First, they are recurrent and relatively close in time to one another (for most groups on a monthly basis). Here, "symbolic" elaboration happens – of beliefs and values about what a "solidarity" purchase is and what it entails, the mission, norms and informal rules of participation: knowledge travels, develops and expands through the network. But the (more or less explicit) assumption is that this is not sufficient: it should be accompanied by a (re)organisation of material flows if any concrete and real change is to happen. Hence meetings also are directed at building relations with the "right" producers, establishing routines and procedural protocols for the group, arranging orders and deliveries, etc. The things and the procedures involved in household shopping (production, delivery, consumption, but also the very bodies of the participants with their wants, dispositions, affordances, tools, etc.) are iteratively shaped in time and space and in this way GAS construct logistics and contents for their shopping that are alternative to those of the large retail organisation.

Symbolic-material construction goes hand in hand with a style of decision-making and action that responds to desire for inclusion, solidarity and equality. For instance, decision-making processes are collegial and thus embody a form of collective-horizontal organisation that is perceived as more resonant to those ideals than a vertical model in which leaders decide for those whom they represent. Every individual takes some responsibility for the GAS's activities – which means that the ongoing maintenance of order is taken directly on themselves by the members of the group and not transferred onto others on other hierarchical scales. The group embodies its own ideals of what a fairer politics is.

### *Re-making responsibility*

The shift from changing lifestyle to changing *life* is quite interesting as a way of comparing the political consumerism of the "consumer-citizen" (Doubleday, 2004) with that of the *gasista*, in particular with reference to the idea of responsibility. Within the mainstream, typically modern, conception of responsible consumption the main actor is the (sovereign, rational) individual who performs choices, being aware of the consequences of his/her action and choosing the "right" course – defined for example on the basis of claims to abstract justice (e.g. all human beings are equal, all should have the same rights). Yet, this conception of responsibility is abstract itself: it does not call into question the space of choice, its constitution, its features, its underlying logics and (ir)rationalities; it provides a view of a disembodied, rational, atomised individual that disregards all the embodied and relational aspects of existence. In so doing, it tends to promote an "ascetic" kind of responsibility, one that functions by subtraction and negation: avoiding what is 'bad' (Dal Gobbo, 2018).

The politics of responsibility within GAS are completely different. This is not to say that there is disregard for other beings (human and non-human) – so much so that one of the main reasons for joining GAS is to support local producers. Nonetheless, more than choosing in a responsible way, *gasistas* construct assemblages that respond to their conception of what is "right". Hence, instead of being accountable for each of their separate choices of consumption, it is the distributed constitution of the material-symbolic space in which they move to put them in the condition of being "response-able" (Haraway, 2016): they can evaluate and experience the consequences of their own actions relationally, for instance by entertaining direct relationships with the producers from whom they buy. This responsibility we might call "positive" since the aim is not to *avoid* what is bad but to *construct* "good" worlds and to make it together with human, but also non-human, beings.

GAS do so by facilitating the circulation of both resources (information, tasks, money and goods) and common interpretations of reality, thus simultaneously providing a framework for collective action and enabling the actual deployment of alternative lifestyles. In so doing, they bring political consumerism beyond so-called "politics of the self" or "life politics" (Giddens, 1991). They do not merely seek to compensate for the inability of institutions to pursue pro-environmental policies and human rights protection. GAS are collective experiences, designed to co-produce the common good by (re)building reciprocity and trust among diverse subjects operating in the same territory, directly intervening in local food provisioning chains, and reintroducing social and environmental sustainability issues in regional economies, sometimes with the explicit aim of participating in the governance of the territory. Hence GAS move beyond political consumerism as a form of mere individual responsibility (Micheletti, 2009) to develop collective, citizenship-driven alternative styles of provisioning. That critical consumption is exercised collectively and face to face has the effect of (re)socialising citizens into many aspects of their consumption, towards what they call "co-production".

GAS thus shape ecologies, assemblages that interweave human relations, worldviews, information, preferences, political demands with the materiality of practices, energies and things: all these aspects take shape and evolve together. We see, thus, a shift from taking "personal responsibility" as an individual to a collective effort for building assemblages of organised co-production. This is reflected in the interesting observation that while for most people GAS's objective is to promote alternative lifestyles, the most commonly perceived outcome is in fact that of supporting local producers (see Figure 5.2). The first aim, somewhat individualised, leaves space to a more directly "political" and collective one: gasistas end up changing the life and work conditions of other fellow beings – the producers – while trying to open a crack into, and building alternatives to, the exploitative and unsustainable mainstream production and distribution system.

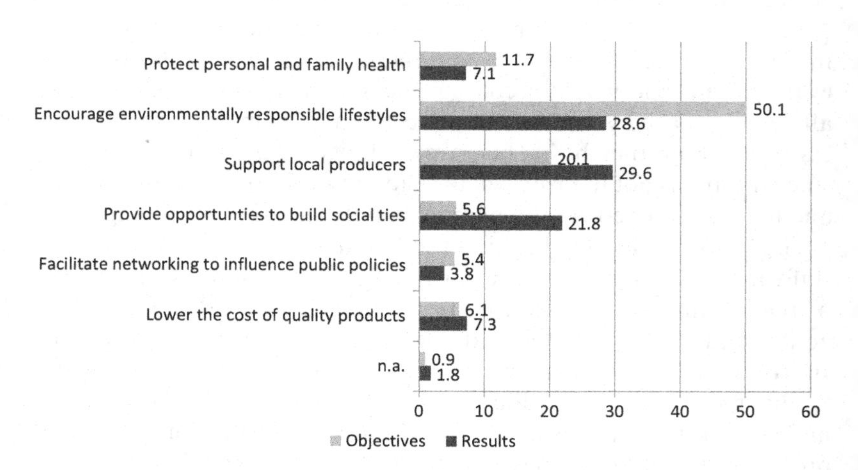


*Figure 5.2* Main objectives and results of GAS groups (%).
Credit: Osservatorio CORES 2013.

***A politics of more-than-human desire***

We might reformulate the argument above by saying that being part of a GAS is one of the ways in which participants pursue happiness: instead of negating themselves something "wrong", they put in place the material conditions that allow them to "live well". It is a desiring kind of politics, but not in the (individualistic) sense of satisfying one's own pleasures or of taming a middle-class consumer guilt for the consequences of one's purchases. Nor, on the contrary, is it a matter of immolating for the other by self-sacrificing. What GAS try to do is building a socio-ecological fabric that, woven together, is able to constitute the possibilities for the beings involved to receive care, be recognised, respected.

GAS are perceived as significantly opening "opportunities to build social ties" (Figure 5.2). Relations are thus very important for the collective pursuit of a "togetherness", which is grounded in the practices of the groups. However seemingly inefficient, time-consuming methods of collective decision-making, home collections, visit to producers, etc. force every member to be proactive, to participate in the group with equal responsibilities and to engage in intensive processes of socialisation (telephone calls and emails, visits, participation to cultural and political events of common interest, the circulation of relevant readings and news and the opening up of space for discussion and exchange). Interestingly, intensifying relations also pushes *gasistas* to engage in politics beyond GAS themselves: they become more collaborative, more interested in (especially local) politics, have an enhanced sense of social efficacy. The pursuit of a better life here shifts the space of political agency from an individualistic (self) imposition of behaviours to a collective construction of new assemblages.

This kind of desiring politics of consumption (and beyond) articulates environmental responsibility and concerns in a very specific way. Ecological sustainability is not at the fore of *gasistas'* concerns. At least initially, engagement passes through the participants' and their family members' bodies, since the main reported reason for participating to GAS is health (see Figure 5.3). Yet, it would be misleading to say that it is self-interest to push *gasistas* to the groups: apart from being health almost as important as "fairness for producers", we have proof of this in the emphasis on *family*, especially children's, health. This embodied pursuit of care for the self and proximate others, by being part of a collective effort of re-shaping trans-local ecologies and socio-ecologic relations, is brought beyond itself to embrace care for distant others (producers) and non-human beings. For instance, in

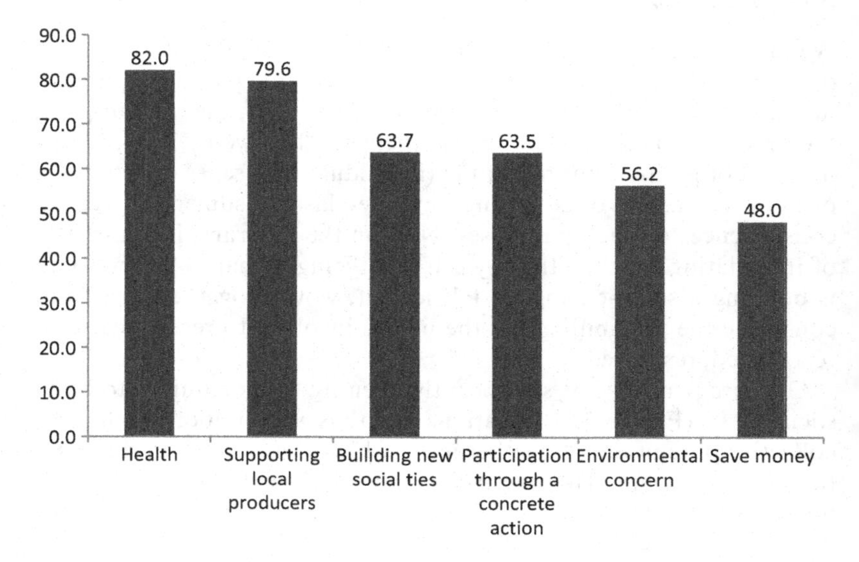


*Figure 5.3* Reasons for joining GAS.
Credit: Osservatorio CORES 2013.

time, consumption of local, organic and seasonal food increases, while the consumption of meat decreases.

In other words, while not looking for an abstract "justice", *gasistas* develop through embodied engagement a care for (local) ecology. They participate to the wider ecologies in which they are embedded: from this very material emplacement awareness grows of the environmental aspects of consumption. This is, as before, also cultivated in the groups through discussions and debates that articulate the diverse ways of intending this emplacement and relationality that involves both humans and non-humans. Care grows from, and grounds in, not principles but the weaving of alliances among human and non-human beings, the evidence of their interdependence emerging from situated experience.

Hence the procedural principle of not creating hierarchies and favouring inclusion and equity extends to the non-human: becoming part of a common enterprise of living together, in which no being is disposable, unnecessary, unimportant. This is also a part of building assemblages: making a world in which different inhabitants might find themselves on a common plane of existence that does not discriminate between what is "higher" or "lower"; the understanding that there is no self-realisation, no enjoyment, at the expense of one's ecologies. Although this is far from being accomplished and straightforward, we might say that as assemblages of food are immanently woven together through matters and words, actions and rhythms, a new kind of horizontal politics starts to emerge: one that not only privileges non-hierarchical relations among human beings but also between them and non-human nature. A bundling and co-emerging where politics, sociality, economy and ecology start to slowly find (back) their unity.

## Conclusion: promises and challenges of SCMOs

GAS can be considered a particular form of SCMOs and, according to our discussion, we argue that they can be considered a fully-fledged political practice. Yet, such practice significantly departs, with the opportunities and challenges that this entails, from what has been defined as "politics" in modern times. Many of the features characterising Modern politics are, in fact, if not absent, at least thoroughly re-articulated by these forms of collective organisation. Instead of vertical models in which many individuals find unity and articulation in institutions that somehow are beyond them; GAS constitute horizontally, they are made of "rhizomatic" (Deleuze and Guattari, 2014), immanent relations that are not reified and fixed into something neatly defined. This is evident in the fact that, even in the face of legislation recognising them as non-profit organisations, 76.1% of GAS tend to remain unregistered and hence to avoid the need of acquiring a formal structure. To the modern rational and disembodied individual who pursues a moral good, GAS's alternative is an embodied, caring and desiring human being who struggles for a common "living well"[1] – for both human and non-human beings (Lambek, 2010). To the politics of the nation-state they oppose one that is trans-local, that horizontally expands by weaving together beings in assemblages of affinity (Torre, 2018). Against abstract choice within a world, a concrete *making* of worlds.

Such prefigurative style of politics holds promises for the transformation of our societies in a more sustainable sense for more than one reason. First, as said, they *embody* instead of merely delineating or asking, examples of different ways of organising and making the world. In so doing, they are not only examples but also reservoirs of developed and working knowledge, capability, skill and relation repertoires that might be of help in the scaling up of what are admittedly niche experiments. They show how it is possible to make politics not separately from the production and reproduction of life, but *in* the very acts of producing and reproducing. Social and ecological sustainability are already present within this framework because they are the premise of the ongoing maintenance of embedded relationalities as constructed around shared ways of feeling and interpreting the world, values of solidarity, mutual help and care.

The danger nonetheless is that these experiences remain insular and live alongside the rest of reality without effectively connecting with it and transforming its wider patterning. The logics of GAS politics do tend in some ways to "spill out": people tend to change other everyday habits (e.g. increased attention to recycling and reduced consumption of electricity and water); participate in other forms of political engagement. GAS themselves promote other solidarity economy activities such as farmers' market, barter groups, etc. Nonetheless, these are at present far from being widespread enough to counterbalance established unsustainable models of production, distribution and consumption, or to provide fully-fledged alternatives to them. Especially as GAS purchase remains a limited aspect of everyday consumption, GAS's greatest significance seems to lay in their being "citizenship labs" more than a real challenge to the global capitalist economy (Forno et al., 2015).

Reticence to organising, expanding and relating to more formal political institutions in order to preserve the direct and controllable ways of coordinating which are felt so important, can be problematic in the face of the huge challenges that the crisis poses us and to which we are called to respond. The latter are seriously global in character and face each localised reality with disproportionate forces. With respect to this, there is a promise in the tendency of GAS and more generally of SCMOs to build networks (e.g. Distretti di Economia Solidale), some of which cross regional and even national borders: they might expand both the significance and the power of these mobilising groups to resist neoliberal orders and provide feasible alternatives (Forno and Maurano, 2016).

Admittedly, as much as the GAS model may have partially succeeded in the past ten years, it is still far from scaling up as a convenient and suitable alternative for most of the population. It might have inspired the (hesitant) appearance of private digital food platforms that aim to shorten the supply chain by intermediating between consumers and producers (see Corvo and Matacena, 2018); furthermore, interest in the creation of local or trans-local ways of food consumption and distribution is giving rise to a number of environmentally and socially conscious initiatives, such as Municipal food policy councils and policies. Yet, grocery shopping is still dominated by mass retailers capable of offering a wider range of goods and foods at lower prices.

At present, becoming a *gasista* involves costs that are insurmountable to most people, and especially to families and individuals at the margins of society. Leaving aside the possibility that ethicality, authenticity and sustainability could just be new ways the bourgeoisie "reinvents distinction" in a period of widespread food commodification (Paddock, 2016), partaking in a GAS is both resources demanding and time consuming. Not only families need to be willing to spend a larger share of their income on food consumption – which by and large explains why it tends to be a middle class practice – but also they need to actively participate in the organisation, share the GAS values and mission, and most importantly guarantee regular attendance so that producers can rely on predictable orders. Clearly, these hidden costs clash with the obstacles that many families, and especially those at the bottom of the income spectrum, face in their lives, which are more and more characterised by job and financial precarity and time constraints. In addition, since the model engages with local production as a way to enhance sustainability, it inevitably tends to neglect the physical and symbolic necessities of the increasing migrant communities that connect to their homes identities precisely by acquiring food and goods from their countries of origin.

Finally, the fact that GAS are partial in their delivery of necessities and that they remain mostly niches poses doubts as to whether their *model* of politics might be altogether ineffective. On the one hand, we might argue that current hegemonic political models have failed to deliver much of their stated claims of justice, equality and now environmental sustainability. It is clear that new models are necessary. On the other, even if the style of politics that GAS embody seems shared by most innovative social movements and mobilisations nowadays, we are still to understand how much this will have significant impacts for social and ecological sustainability in the future. We might say that a new paradigm is slowly delineating, though it needs to overcome the fragmentation of experiences so that the localised embeddedness that constitutes the force of these movements might not become a particularistic fetter in building alliances for producing significant change in our multiple worlds.

## Note

1 This is reconducible to the feminist critique of the "male" character of Modern politics, so much so that we might argue that the "individual" we are talking about is the white-male-European-heterosexual-rational individual that has imposed himself on the rest of the world as neutral model of human being since the 17th century. In this respect, it is interesting to notice that people taking part to GAS are overwhelmingly women.

## Bibliography

Asara, Viviana, Otero, Iago, Demaria, Federico, & Corbera, Esteve. 2015. Special feature: Socially sustainable degrowth as a social-ecological transformation. *Sustainability Science*, 10(3), 375–384.

Bosi, Lorenzo, and Zamponi, Lorenzo. 2015. Direct social actions and economic crises: The relationship between forms of action and socio-economic context in Italy. *Partecipazione e Conflitto*, 8(2), 367–391.

Brunori, G., Rossi, A., & Guidi, F. 2012. On the new social relations around and beyond food. Analysing consumers' role and action in Gruppi di Acquisto Solidale (Solidarity Purchasing Groups). *Sociologia Ruralis*, 52(1), 1–30.

Castells, Manuel, Caraca, Joao, & Cardoso, Gustavo. (Eds.). 2012. *Aftermath: The Cultures of the Economic Crisis*. Oxford: Oxford University Press.

Corvo, Paolo, & Matacena, Raffaele. 2018. The new 'online' alternative food networks as a socio-technical innovation in the local food economy". In: Scupola, Ada & Fuglsang, Lars (Eds.) *Services, Experiences and Innovation. Integrating and Extending Research*. Cheltenham: Edward Elgar Publishing, 301–315.

D'Alisa, Giacomo, Forno, Francesca, & Maurano, Simon. 2015. Grassroots (economic) activism in times of crisis: Mapping the redundancy of collective actions. *Partecipazione & Conflitto*, 8(2), 328–342.

Dal Gobbo, A. 2018. Desiring ethics: Reflections on veganism from an observational study of transitions in everyday energy use. *Relations*, 6(2), 233–250.

Dal Gobbo, Alice, & Torre, Salvo. 2019. Lavoro, natura e valore: logiche di sfruttamento, politiche del vivente. *Etica e Politica*, 21, 165–171.

Deleuze, Gilles, & Guattari, Felix. 2014. *A Thousand Plateaus. Capitalism and Schizophrenia*. London: Bloomsbury Academic.

della Porta, Donatella. 2006. *Globalization from Below: Transnational Activists and Protest Networks*. Minneapolis: University of Minnesota Press.

Doubleday, Robert. 2004. Institutionalising non-governmental organisation dialogue at Unilever: Framing the public as 'consumer-citizens'. *Science and Public Policy*, 31(2), 117–126.

Forno, Francesca, Grasseni, Cristina, & Signori, Silvana. 2015. Italy's solidarity purchase groups as 'citizenship labs.' In: Kennedy, Emily Huddart, Cohen, Maurie J., Krogman, Naomi & Sustainable Consumption Research and Action Initiative (Eds.), *Putting Sustainability into Practice: Applications and Advances in Research on Sustainable Consumption*. Cheltenham: Edward Elgar Publishing, 67–90.

Forno, Francesca, & Graziano, Paolo R. 2014. Sustainable community movement organisations. *Journal of Consumer Culture*, 14(2), 139–157.

Forno, Francesca, & Graziano, Paolo. 2019. From global to glocal. Sustainable community movement organisations (SCMOs) in times of crisis. *European Societies*, 21(5), 1–24.

Forno, Francesca, & Gunnarson, Carina. 2011. Everyday shopping to fight the Mafia in Italy. In: Micheletti, Michele, & McFarland, Andrew S. (Eds.),

*Creative Participation: Responsibility- Taking in the Political World*. London and Boulder, CO: Paradigm, 103–126.

Forno, Francesca, & Maurano, Simon. 2016. Cibo, Sostenibilità e Territorio. Dai Sistemi Di Approvvigionamento Alternativi Ai Food Policy Councils. *Rivista Geografica Italiana*, 123, 1–20.

Giddens, A. 1991. *Modernity and Self-identity: Self and Society in the Late Modern Age*. Stanford, CA: Stanford University Press.

Giugni, Marco, & Grasso, Maria T. (Eds.) 2018. *Citizens and the Crisis*. London: Palgrave.

Grasseni, Cristina. 2014. Seeds of trust. Italy's Gruppi Di Acquisto Solidale (Solidarity Purchase Groups). *Journal of Political Ecology*, 21, 127–221.

Graziano, Paolo R., & Forno, Francesca. 2012. Political consumerism and new forms of political participation: The Gruppi Di Acquisto Solidale in Italy. *The ANNALS of the American Academy of Political and Social Science*, 644(1), 121–133.

Guidi, Riccardo, & Andretta, Massimiliano. 2015. Between resistance and resilience: How do Italian solidarity based purchase groups change in times of crisis and austerity? *Partecipazione e confitto*, 8(2), 443–447.

Haraway, Donna, J. 2016. *Staying with the Trouble: Making Kin in the Chthulucene*. Durham, NC: Duke University Press.

Harvey, David. 2007. *A Brief History of Neoliberalism*. New York: Oxford University Press.

Kousis, Maria. 2017. Alternative forms of resilience confronting hard economic times: A South European perspective. *Partecipazione e Conflitto*, 10(1), 119–135.

Lambek, M. 2010. *Ordinary Ethics. Anthropology, Language, and Action*. New York: Fordham University Press.

Lefebvre, Henri. 2014. *The Critique of Everyday Life*. London: Verso.

Leonardi, Emanuele. 2019. Bringing class analysis back in: Assessing the transformation of the value-nature nexus to strengthen the connection between degrowth and environmental justice. *Ecological Economics*, 156, 83–90.

Matacena, Raffaele. 2016. Linking alternative food networks and urban food policy: A step forward in the transition towards a sustainable and equitable food system? *International Review of Social Research*, 6(1), 49–58.

Meyer, David, & Whittier, Nancy. 1994. Social movements spill-over. *Social Problems*, 42(2), 277–297.

Micheletti, Michele. 2003. *Political Virtue and Shopping: Individuals, Consumerism and Collective Action*. London: Palgrave Macmillan.

Micheletti, Michele. 2009. La svolta dei consumatori nella responsabilità e nella cittadinanza. *Partecipazione e Conflitto*, 3, 17–41.

Micheletti, Michele, & McFarland, Andrew. (Eds.) 2011. *Creative Participation: Responsibility Taking in the Political World*. Boulder, CO and London: Paradigm.

Paddock, Jessica. 2016. Positioning food cultures: 'Alternative' food as distinctive consumer practice. *Sociology*, 50(6), 1039–1055.

Schlosberg, David, & Coles, Roman. 2016. The new environmentalism of everyday life: Sustainability, material flows and movements. *Contemporary Political Theory*, 15, 160–181.

Signori, Silvana, & Francesca Forno. 2019. Consumer groups as grassroots social innovation niches. *British Food Journal*, 121(3), 803–814.

Torre, Salvo. 2018. *Contro la frammentazione: movimenti sociali e spazio della politica*. Verona: Ombre corte.

# 6 Black diaspora women lead cooperative banks[1]

*Caroline Shenaz Hossein*

## Introduction

Women around the world participate in informal cooperative banks. In January 2019, one million women in Kerala, India, organized themselves in a Wall of Protest (called *vanitha mathil* in the local language of Malayalam) to speak up against unjust political matters and they were able to do this through their relationship in the Kudumbashree movement, a cooperative-like institution building up ones economic power (Thiagarajan, 2019; Christabell, 2013; Datta).[2] In Niger, tens of thousands of women are part of the Mata Masu Dubara movement, a system of self-managed village banks (Grant & Allen, 2002). Informal banking cooperatives are institutions where people collectively lend and save amongst their peers (Chiteji, 2002; Rutherford, 2000; Ardener & Burman, 1996; Geertz, 1962).[3]

And Black women around the world are standing up and organizing cooperative banks to help themselves and one another

> Ladies, please jus' fix your life and don't wait around…do something for you, your life and your children's lives. Join our susu. What will the result be? Happy women.
>
> ("Mabinty," Sierra Leonean-Canadian woman, 20 March 2015)

Women like "Mabinty" in the quote above create self-managed banks in their communities that focus on female needs. Ardener and Burman (1996), in their classic text *Money-Go-Rounds: The Importance of Rotating Savings and Credit Associations for Women,* document the phenomenon of informal cooperative banks, also known as rotating savings and credit associations, are grass-roots financial collectives.

In this study, I argue that major cities in the Americas, specifically in Canada and the Caribbean abound with credit unions and commercial banks, that Black diaspora women organize cooperative banks because they feel unwelcomed in commercial banks and in doing so they show that they can do business differently in society. These women show that such collectives are a source of comradery as well as a process of building financial capabilities that have always been around.

So, when the financial crises of 2007/2008 in the US and the Greek banking crisis of 2015 occurred, Black women remained committed to their informal cooperative. A scholar of community economies, Stephen Healy (2009) holds that formal commercial banks have left most people disillusioned about their ethics. British people ignited a movement called People to People Finance Association as a way for citizens to self-manage their own monies. Yet Black women in the Americas who have had generations of distrust of conventional banks have stayed under the radar in terms of their organizing of alternative cooperative banks, with no real intention to develop what they do into a credit union or a commercial bank (Hossein, 2018a; Ardener & Burman, 1996).

Because these cooperative banks are embedded in social relationships and considerate of people's social lives, it has been important that they be informal to assist those most injured by the wicked racism in society against non-white minorities. It should be emphasized from the outset that these cooperative banks are not a premature form of credit unions; rather, they are entities in their own right. In this chapter, I argue that women who organize informal cooperative banks—referred to locally as *susu*, *partner*, *meeting-turn*, *box-hand*, *sol* and many other names—in order find inclusive and community focused economies because commercial banks are too individualistic, exclusionary and elitist.

## Women's contributions to cooperative banking

The story of cooperative banks usually starts in the 1880s with the German Raifeissen's cooperative banks or the Rochdale weavers story (Guinnane, 2001; Fairbairn, 1994). In Canada, the French-Canadian Desjardins' *caisses populaires* of the 1900s dominate the discourse about cooperative banking in the Americas (Mendell, 2009). Black women, as well as many other racialized women in the Americas have also engaged in cooperative banks for a very long time—at least as far back as the 1600s—long before such institutions were named (Hossein, 2018a, 2013). Gordon Nembhard (2014) traces self-help groups of African-Americans to the 1700s, when people created cooperatives, mostly informal ones, as a way to resist racism. What is certain is that collective banking has been ingrained in the Black diaspora's DNA out of survival and with time it has been held as a sacred tradition for many people.

Cooperative banks, mostly informal, are a deeply held African tradition. It also speaks to the functionality of "getting things done" by a historically oppressed group of people. These self-help groups are THE REAL LIFE aspect of the social economy, in which women come together to redo financing systems. In this study, women reported that these banks are vital forms of support for them in terms of their livelihoods needs but also their social networks (Hossein 2013). As "Betty," a Jamaican-Canadian single mother of five children in Toronto, explained,

> Oh yeah, I used to run money for my mother back in Jamaica. I used to see how them women used to uplift one another with partner [an informal cooperative bank]. It was a serious, serious time when money time and this instilled in me how to work with people I know to get what we need. I am not waiting on outdoor persons (referring to banks and others from outside the community) to do that for me.
>
> (31 July 2015)

"Betty" thus knew from a very young age that women coming together to discuss money was "serious business." When "Betty" emigrated to Canada, *partner* was the lifeline that helped her build her investments, start a business, attend college and buy a house.

While the literature on informal banks is extensive, and examines people's ingenuity in creating local banks, it does not discuss the agency of uneducated Black women in organizing cooperatives. Informal cooperative banks are unregulated financial groups that provide quick access to savings and credit for people, mostly women, who belong to the same socio-economic groups (Hossein, 2017; Rutherford, 2000; Ardener & Burman, 1996; Geertz, 1962). I first show Black women building cooperative banks. Second, I outline my empirical methods as well my theoretical influences in examining informal cooperative banks in the Caribbean and Canada. Finally, I conclude with the findings that emphasize that Black women have contributed to building sustainable communities.

## Brief historical account of informal cooperative banks in the Black Americas

The African diaspora in the Caribbean and in Canada has been deeply affected by enslavement, colonization and racism (Benjamin & Hall, 2010; James et al., 2010; Mensah, 2010; James, 1989). It is during these critical moments in history that persons of African descent have rethought how to organize their social and business lives. Africans and Caribbean people have embraced the informality of these systems. In the Canadian case and each of the Caribbean cases—Jamaica, Haiti, Trinidad and Guyana—African slaves and their descendants carried out market activities and engaged in informal money clubs (St. Pierre, 1999; N'Zengou-Tayo, 1998; Harrison, 1998; Wong, 1996; Witter, 1989; Mintz, 1955). In the Americas, the informal cooperative banks of today are a deeply valued African tradition rooted in ancient systems of *susus* and *tontines* brought over by African slaves (Hossein, 2014a; Heinl & Heinl, 2005; St. Pierre, 1999; Mintz, 1955).

African slaves in the Americas expressed their defiance of slavery when they pooled their earnings from the market and rotated lump sums of money to each other without their masters' permission (St. Pierre, 1999). Faye V. Harrison's work (1988) shows that since slavery, Jamaican women vendors used partner to meet their livelihood needs. In Toronto, Canada, the women I interviewed stated that informal cooperative banks are part of the many financial devices they draw on. Women also made it clear that informal cooperative banks help them to avoid using pay-day lenders and their exorbitant fees (Jane and Finch, 20 March 2015).

Under colonization, banks did not lend to the local Black population, and especially not to women, so women turned to the community associations handed down to them by the generations before them. In Haiti many women took risks, as it was illegal to organize in groups. The documentary *Poto Mitan: Haitian Women, Pillars of a Global Economy* (2009) shows how women in Cité Soleil (a slum in the capital city) rejected low-paid factory work and turn to *sol* (a money pool) to start their own businesses. Such banks are not only in Global South poor areas but are very much a part of the life in cities around the world (Hossein, 2017; Ardener & Burman, 1996). Women often hide their money from male partners (Mayoux, 1999; Nelson, 1996). In a focus group in Toronto, for example, a 40-year-old Guyanese-born Canadian woman made this point to the group of women (who were all nodding in agreement):

> So yes, I hide it [money from the cooperative] from him [her spouse]. I don't know what he is doing and I am protecting myself. I don't know the thoughts in his head so I don't care that I hide it from him. One day he can take up and leave me with not a thing.
>
> (Jane and Finch, 20 March 2015)

## Women deliver mutual aid to racially excluded women

Black women in the diaspora have created sustainable money groups for themselves and their communities, despite living in inhospitable environments for centuries. Studies by Gordon Nembhard (2014), Mintz (2010), St. Pierre (1999) and Du Bois (1907) show that Black women in the Americas participated in mutuals and collectives to counteract social exclusion. For centuries, Black women in the Caribbean and Canada have mobilized scarce funds in a collective manner in their low-income communities (Hossein, 2020, 2018a, 2014a; Ardener & Burman, 1996; Niger-Thomas, 1996). Mintz (2010, 1955) showed how the Haitian *madam saras* relied on a *system pratik* and collectives called *sol* to business in the markets. W.E.B Du Bois (1907) recognized very early the work of Black women in group economics both in Africa and the US. Gordon Nembhard (2014) shows the quiet ways Black women organize to protest and to feed their families and communities. African American economist Ngina Chiteji (2002) argues that informal cooperative banks should be counted as part of the financial landscape because they are cost effective and useful to people cut out of formal banks.

Places outside of the Americas, governments have acknowledged the relevance of informal cooperatives in the economy. In a recent trip to Ghana in 2017, one learns quite fast that there formal financial system includes an array of providers such as commercial banks, credit unions, microfinance banks as well as Susu enterprises, which are akin to cooperatives in this paper. The Indian state found the reliability of informal banks to be valued and created laws as far back as the Travancore Chitties Act of 1918 and Madras, the Cochin Kries Act in 1931 and the Chit Fund Act of 1961 to recognize these mostly women-run cooperatives (see Sethi, 1996: p. 171 for these acts). Most of India is governed by the Chit Funds Act 1982 (Act No. 40 of 1982) to ensure protection for the millions that engage in chit funds (Sethi, 1996: p. 172).

## Black political economy to understand diaspora lives

At least 200 million persons of African descent live in the Americas. And it seems very fitting—in the U.N. Decade of the Year of Persons of African Descent 2015–2024—to examine one of the world's most used interventions, informal cooperative banks. Far too many social movements advocating for political change often leave out the economic and business aspect. My work is unearthing – still developing – this idea of a Black Social Economy which is politicized for goodness in that it can coopt business in ways that work for local people (Hossein, 2019, 2018a).

Cooperative economics which is informal and focused on racial minorities benefits from drawing on the theorizing of Black thinkers—such as W.E.B Du Bois, Booker T and the ground knowledge of the Underground Railroad—who focus on self-reliance, group economics and alternative economics as a pathway out of a racist society. I credit the Underground Railroad as an important form on cooperation that had to be hidden because Tubman and the abolitionists had to bring hundreds of slaves into Canada through intricate cooperative systems (James et al., 2010). The concept of economic cooperative is at the very core to understanding what it means to have Black economic livelihoods in the North American context.

Mutual aid was part of survival for enslaved and colonized Black people in the US, Canada and the Caribbean. In Washington's seminal work, *Up from Slavery* (first printed in 1901), he supported Black entrepreneurialism as a way to lead to mutual progress for an excluded group of people. Washington attracted criticism for his accommodating views on industrial trades; however, it must be remembered that Washington, born into slavery, was committed to the common cause to end violence against African-Americans and he used his money to fund anti-lynching groups, and he did this collectively. His ideas of self-help were of great service to many people engaged in Garvey's UNIA movement, which was collective and member-based. As early as 1903, African American and Harvard-educated W.E.B Du Bois advanced the theory of group economics among Black people to withstand white racist power. Du Bois' powerful piece *The Souls of Black Folks* (2007/1903) describes communal and collective forms of African business, and this historical grounding is inspiring for Black people who live outside of the African continent.

## Methods

This study draws on multiple qualitative methods to research the attitudes of women who participate in cooperative banks which are informal in 16 low-income communities in five Caribbean countries and in Canada's largest city Toronto. As shown in Table 6.1, I interviewed individually and held several focus groups with 332 people in Jamaica, Guyana, Trinidad and Tobago, Haiti and Canada from June 2007 to July 2015.[4] I also carried out in-depth interviews over a five-month period with three women in charge of informal banks in Kingston, Jamaica.

*Table 6.1* Interviewed Black women in informal cooperative banks in the Caribbean and Canada

| *Country/Method* | *Jamaica* | *Guyana* | *Haiti* | *Trinidad* | *Canada* | *Total* |
|---|---|---|---|---|---|---|
| Number of women in informal banks, from focus groups | 57 | 5 | 74 | 0 | 46 | 182 |
| Individual interviews with women in informal banks | 89 | 14 | 19 | 23 | 5 | 150 |
| Total | 146 | 19 | 93 | 23 | 51 | 332 |

Source: Author's data collection from 2007 to 2017.

In Kingston, I interviewed women in the downtown communities south of Cross Roads, including the neighborhoods of Trench Town, Bennett's Land, Whitfield Town, Rosetown, and Frog City, and the former prime minister's constituency of Denham Town and Tivoli Gardens in 2009. In Haiti, I interviewed women in Cité Soleil, Carrefour, Martissant, and La Saline, as well as Bel Air in Centre-Ville and Jalousie and Flipo in the hills of Pétion-Ville. The focus groups in Haiti were held in the poor areas of Bon Repos, Port-au-Prince, between the periods of 2009 and 2012. In 2008 and 2010, I interviewed women in Albouystown in Georgetown, which is ethnically diverse and has a large Afro-Guyanese population, dougla (mixed race of African and Indian background) population, as well as East Indian, Portuguese and Amerindian people. In 2013, I interviewed women in Laventille, Beetham Gardens and Sea Lots in east Port-of-Spain, Trinidad. In 2015 to 2017, I carried out focus group sessions with 46 women in the Jane and Finch and southwest Scarborough communities in Toronto, Canada.

In the focus group sessions in the Caribbean, I asked women, "What kind of financial provider meets the needs of persons in poor communities?" and soon realized that informal cooperative banks were the most prominent financial device they used. Once I was aware of the relevance of informal cooperative banks, I followed up with questions such as, "With many banking options close by, why are informal banks so prevalent?" and "Why do persons organize and join Informal cooperative banks (money groups)?"

## Findings: Black women build cooperative banks

Black women have a profound influence on alternative economics when they organize informal cooperative banks to increase the financial options of excluded groups. Not only do informal cooperative banks give women choices over where to bank, but these groups also restore women's faith in banking after enduring discrimination and humiliation in their everyday lives. In this findings section, I show that Black women in five countries, in Canada and the Caribbean, are at the forefront of cooperative development.[5]

### *Informal cooperative banks in Guyana and Trinidad*

African slaves brought with them West African traditions of *susu*, through which they mobilized savings (and loaned out money) on a weekly basis (Mintz 2010, 1955; St. Pierre 1999). Even under slavery or indentured servitude, Africans carried out sideline businesses and held market days with the extra provisions they grew. After slavery was abolished, the British colonialists imported indentured servants from India to Trinidad and Tobago and Guyana. Africans were free, but the bankers and planters made it difficult for them to conduct business. To counteract exclusion in business, Africans pooled resources in money clubs to buy plots of lands and villages.

The structure of *susus* varies from community to community, but members are usually self-selected by individuals who know each other, and they determine the fixed deposit they will contribute every week. The informal banker who runs a business out of her home allows members to pass by to drop off their deposits. The group usually decides how long they will do this for, but commonly it is a period of 10 to 12 months. Once all the members agree on the rules and structure of the *susu*, then the head person in charge of the bank launches the cooperative with the first in-take of deposits. Women in charge of these banks claim that they lend out those deposits to the members the same day they take in the money in order to avoid having large sums of cash on their person. The system of rotation can take a number of forms, which again varies based on the group dynamics. Money can be allocated based on first-come, first-serve; need (i.e., seasonal work, funds tied to business needs, personal crisis); or lottery (i.e., drawing names).

The culturally distinct lenders in Jamaica, Trinidad and Guyana have a tense interaction with borrowers who differ from them in terms of class, culture and sometimes gender. Class-based racism and partisan politics in Trinidad and Guyana have interfered with people's access to finance (see more in Hossein 2015, 2014a). As of April 2015, in both Trinidad and Tobago and Guyana, Indo-Caribbean political leaders dominate national politics to the exclusion of Afro-Caribbean people. A pervasive cultural narrative disparages the business acumen of African-descendants and the commercial bankers, usually educated men of East Indian descent, are hesitant to make loans to poor Black people (Hossein 2015, 2014a). In Guyana's microfinance sector, the main specialized microfinance agencies are managed and staffed by educated middle-class Indo-Guyanese who lend to Indo-Guyanese clients (Hossein 2014a). In my study, at least 65% of the entrepreneurs interviewed in Allbouys town claimed that they borrowed money from box-hand bank because they could not access a bank loan. Black business people, mostly women, in Guyana and Trinidad cannot easily access bank loans because of gender, identity and party politics, and they inevitably turn to cooperatives to meet their business needs (Hossein, 2015, 2014b).

### *Haiti's cooperative history*

Haiti's cooperative development has been exceptional. In former colonies around the world, cooperatives have been the project of local or foreign political elites, thus creating top-down control and limited development (Develtere, 1993). Yet in Haiti cooperative development was inherited by a cultural tradition of pooling money by African slaves when they first arrived in the 1600s (Mintz, 2010). In French-speaking Benin and Togo, West African countries that Haitians claim as their ancestral lands have strong traditions of *tontines* (informal cooperative bank). *Sols* in Haiti also reach millions of people. The first Haitian cooperative was formalized in 1937 in Port-a-Piment du Nord, near Gonaïves, soon after the US occupation ended (Montasse, 1983). More *caisses populaires* (credit unions) were formalized in La Valée (Jacmel) in 1946 and in Cavaillon (South) and Sainte Anne in Port-au-Prince in 1951, during a time of repressive politics (ibid.).

Despite Haitians being banned from *gwoupmans* (a local term referring to associations) and cooperatives under the brutal Duvalier dictatorships (1957–1986) of Francois "Papa Doc" and Jean-Claude "Baby Doc Duvalier (N'Zengou-Tayo, 1998), the masses participated in *sols*, cooperatives and *caisses populaires* to meet their needs (Maguire, 1997). Haitian cooperative scholar Emmanuel Montasse (1983) discovered a growth of credit unions in the period from 1951 to 1983, and suggests that it occurred because during these years people were deprived of basic services.

African ideas of *kombit* (collectivity) come from the Beninese (then Dahomey) ancestors who brought banking concepts to the Americas as far back as the 1600s when slavery started in Santo Domingo (then Saint Domingue, now Haiti). The country's politics has been oppressive since independence, and leaders since Jean-Jacques Dessalines (1804–1806) have adhered to politiques *du ventre* (politics of the belly) dictatorships, leaving the masses in complete suffering. The first formal financial *caisses populaires* created in 1937 in Port-a-Piment du Nord, near Gonaïve, was no doubt cultivated by the traditions of sols. At least 80% of the 10 million Haitians rely on sols to meet their everyday financial needs. The local traditions of *kombit*, *gwoupmans* and *sols* were ways for excluded Haitians to create civil society groups—and they are testimony to the democratic spirit of the masses (Fatton, 2007; Montasse, 1983). One banker interviewed attested to the importance of *sols* in Haitian society:

> Caisses populaires belong to the Haiti people. These caisses are accessible, grassroots and embedded into people's hearts, because they focus on people's community, collectivity, and helping each other out which are very important traits for us [Haitians] especially those of us who are poor.
>
> (Senior banker, 2 October 2010)

Sols are trusted by people in the community. Every month or week, members contribute a fixed amount, such as 100 gourdes (about US $2.5), for a cycle that can range from six to ten months, depending on the number of members. Members agree to contribute regular savings, and when their turn comes, they can use the money for a specified period, as managed by the banker, the "Mama Sol," who is usually uneducated. People explained to me that this system creates a place for the poor to save and borrow money. Sols may be completely free with no fixed fees, or may apply a small flat fee for the duration of the membership (Bon Repos, 9 October 2010). Sols are low cost and trusted by their users because of their grass roots and collective nature. Furthermore, this capital, mobilized from the grass-roots, contributes to local organizing and brings people who are normally ignored to feel a part of their community (Bon Repos, 9 October 2010).

### *Jamaican women and partner banks*

The Jamaican *partner* is a home-grown institution for local people of all classes, and it is a life-line for those who have no other banking alternative (informal bankers, March to July 2009). The cultural context helps to explain why *partner* banks are so relevant in Jamaican society. Politics in Kingston, Jamaica's main urban center, are marred at election times by violence; and political elites, usually the ones who have power, make promises of money, lodgings and jobs to very poor (dark-skinned) political activists. If they fail to deliver the vote for their candidate, they lose the political hand-outs. Academics have written extensively on this entrenched mechanism, in which elites use uneducated masses in the downtown slums to carry out heinous crimes to assure votes and political victory in exchange for housing or other financial benefits (Tafari-Ama, 2019; Sives, 2010). Years of politicians and gangsters using downtown residents to carry out their criminal work has led people to distrust the political and business elites (Hossein, 2016; ibid.).

Much of the attraction of *partner* lies in the fact that these institutions are run by ordinary women who know the day-to-day reality of the people in their communities. Social exclusion from commercial banks has driven up the demand for informal banks (Hossein, 2015, 2014a), and so too has the need for individuals to rely less on unscrupulous lenders—loans that would increase ties to political elites or informal leaders. Tucked away behind her metal cage, "Rickie," a 29-year-old bar owner, was thankful for me asking about Jamaican partner banks:

> Pardna. Live for dat ting. Most people here [in this low income community] don't have go to banks. Dem [the bankers] don't know what's going on here and wi na know what's going on in their banks. Downtown know Pardna … it is the one ting here for wi.
>
> (9 June 2009)

Jamaican political scientist Obika Gray (2004) similarly points to the widespread urban resistance as "social power" among the urban poor, including among small businesses. Across the Caribbean region, members of the African diaspora turn to local informal financial groups that they know and trust as a way to harness their own power and to rethink the financial institutions they want in their lives. A member of an informal cooperative bank, called Jamaican Partner, who wished to be anonymous stated,

> Partna is fi wi, and bank is fi di big man uptown—that is, the partner bank is for the poor [us] and formal banks are for the rich. Yuh don't have to be rich or educated to throw partna.
>
> (July 2009)

Economists Handa and Kirton (1999) surveyed 1,000 people in Kingston, and found that 75% of the informal bank users were women between the ages of 26 and 35 who organized partner for an average of nine years. These are people who are aware of the community's needs. For example, "Miss Paddy" has never held a bank account at a commercial bank or credit union and used informal cooperative banks for her banking needs: She is one of the thousands of Jamaicans living in downtown communities who do not have the birth certificate required to open an account (6 May 2009).

Women in charge of partner banks are not trained as bankers, yet they manage significant sums of monies like trained bankers. They decide who gets access to the lump sum first, and they assess the person's risks for defaulting, as a trained loans officer would do. The sustainability of these systems proves their viability. Partner banks are made up of a group of people who know each other, sometimes they are related (Three informal bankers, March to July 2009). Several variants of the partner bank exist, and although all are saving plans, many are also lending plans (Ibid.; Handa & Kirton, 1999). Each person's contribution to the partner bank is called a "hand" and it is "thrown" (deposited) for a designated period of time; the pooled money is called a "draw." Peer dynamics ensure people comply with payment rules, and social sanctions are applied in the case of default.

Jamaican people want financial systems that enable them to do what they need to do without restricting their freedoms and 82% (191 out of 233) of the entrepreneurs I interviewed "throw pardna" (participate in partner). In interviews, partner was the lending model that most people (57%, 133 out of 233) trust and claim meets their needs (and others preferred a variety of different lenders from family, banks, nonprofits, credit unions). Commercial banks ranked fairly low in my study because the model does not seem to reach the women who need them. Women members interviewed said they preferred the partner banks because there was "no rigmarole" (paper work) and there are few fees and easy access (Three informal bankers, March to July 2009). The women interviewed claim that repayment rates are high (usually 100%) in such cooperative banks because people trust these systems.

### *Toronto's cooperative banks rooted in local communities*

About 3% (or 842,000) of Canadians are unbanked (Buckland, 2012) with no access to financial services.[6] This has given rise to studies on alternative banks, such as payday lenders, and microfinance organizations, not much on informal cooperative banks. In *Hard Choices*, Jerry Buckland (2012) examines a number of alternative financial providers in the inner-cities and reasons why excluded people turn to payday lenders for monies. In February 2020, an Egyptian-Canadian, Dana Ramadan, went into her bank, the Royal Bank of Canada in Toronto's west end where she was a customer and her treatment was by tellers was hostile because she wanted to withdraw her money and they reacted harshly by suspecting she was doing something illegal (Paradkar, *Toronto Star* 2020). This case study shows how racialized people will create their own cooperative banks located in their communities with people they know and trust because of the humiliation they endure in commercial banks. A Somali-Canadian resident of Jane and Finch explained:

> We are bringing change through our own banking ways. It helps me to keep doing what I do when they (critics who do not know about informal cooperatives) look at us in a weird way to say it's our culture. We show we can do it [business] ourselves and also … is a social thing … We drink tea and talk.
>
> ("Fardowsa," 20 March 2015)

Some commercial banks seemingly appear to be elitist and biased against certain groups. Again in December 2014, Haitian Canadian Frantz St. Fleur was arrested on the suspicion that he was trying to cash a fraudulent check written by Remax at his Scotiabank branch in Toronto's east end where he was a customer for ten years (*Toronto Star* 2014). In January 2020, Maxwell Johnson, an Indigenous grandfather, and his 12-year-old grand-daughter were handcuffed by police after the Bank of Montreal (BMO) and wrongfully suspected of illicit behavior (Sterritt, CBC 2020). All he wanted was to open a bank account for his grand-daughter for the first time. Many racialized people, and Black-Canadian women in particular, who cannot access loans prefer to start up their own cooperative institutions as "Fardowsa" did in the quote above. This kind of overt humiliation, along with the downsizing of bank branches, has increased the appeal of informal cooperatives (Hossein, 2018b).

Cooperative banks that are informally organized among Toronto's diaspora are diverse with women representing 15 countries in Africa and the Caribbean. Newcomers to Canada stated that informal cooperative banks gave them a sense of community. As "Natla," a 35-year-old married Sudanese-Canadian woman, told me:

> Who knows me here when I first come from Sudan [pause] No one. I can't even speak English back then. Sandooq [an informal cooperative bank] give me friends and a chance. I buy [bought]

> my airplane ticket back home and bring my children for there [for vacation]. Sandooq helped me so much when I came to Canada. I swear to my God for it.
>
> (26 March 2015)

## Conclusion: Black women create sustainable cooperative banks

Millions of people, especially women of the African diaspora engage in cooperative banking. What they do is often obscured, on purpose and/or they are simply ignored with no support from formal cooperative associations or credit unions. The cooperative movement is missing on African diaspora spreading the benefits of a cooperative model for racial justice. Business and social exclusion in conventional banks have made Black women rethink how and where they want to do business. Hundreds of Black women in this study made it clear that they cooperate with one another to achieve their life goals, and in my view, they are consciously taking charge by designing financial systems that are inclusive. Caribbean bankers are familiar with people's affinity for informal cooperatives that commercial banks such as the Bank of Nova Scotia of Jamaica, Jamaica National Building Society and Haiti's Sogebank have offered plans based loosely on these institutions, such as "Partner plan" and "Mama Sol," as a way to appeal to women. In no way do these programs offer the same kind of refuge, but bankers there try to imitate informal banks. This is not the case in Canada. Black Canadian women operate their informal banks out of the public view for fear of being arrested and this requires further investigation why they feel a need to hide their community building.

Black diaspora women in the Americas offer a distinct people-focused form of banking because they understand first-hand the imprint of slavery, colonization and racism has had on Black people. African-inspired informal cooperative banks may not change a woman's social status per se, but they do help mitigate the hostile environments in which they must cope. Black women are able to take a stand, and turn into an expert on finance, just like that, because they are bonded and supported by members who believe in helping each other. These Black women—whether from the Caribbean or Canada, or from low or middle income backgrounds—are proving to be the very epitome of trustworthiness and cooperation, and this may be a major gift to the cooperative movement once they figure out ways how to extend their education and support to these types of groups. Black women are making major contributions to cooperative banking.

## Notes

1 This chapter is a slightly adapted version from the originally published paper in 2015 titled "Black women as cooperators: Rotating savings and credit associations (Informal cooperative banks) in the Caribbean and Canada" in the *Journal of Co-operative Studies*, 48(3): 7–18 and the author was granted permission by the journal editor Jan Myers on 27 March 2019.
2 The Kudumbashree movement has impacted more than five million people. See more at: www.thebetterindia.com/119677/kudumbashree-poverty-gender-5-million-kerala/
3 I use the term informal cooperative banks but also use the terms, financial collectives, ROSCAs and self-managed banks to refer to the same phenomenon.
4 On 12 January 2010, Haiti experienced a 7.0 magnitude earthquake that left 300,000 persons dead and 1.5 million people displaced and living in tent cities.
5 The Canada and Trinidad cases are part of 2013 to 2017 fieldwork for a number of various conferences and papers.
6 Buckland (2012) makes a distinction between unbanked and underbanked.

## Bibliography

Ardener, Shirley, and Sandra Burman. (1996) *Money-Go-Rounds: The Importance of Rotating Savings and Credit Associations for Women.* Oxford, UK: Berg.

Benjamin, Russell, and Gregory Hall. (2010) *Eternal Colonialism.* Lanham, MD: University Press of America.

Buckland, Jerry. (2012) *Hard Choices: Financial Exclusion, Fringe Banks, and Poverty in Urban Canada.* Toronto, ON: University of Toronto Press.

Chiteji, Ngina, S. (2002) "Promises Kept: Enforcement and the Role of Rotating Savings and Credit Associations in an Economy." *Journal of International Development* 14(1): 393–411.

Christabell, P.J. (2013) "Social Innovation for Women Empowerment: Kudumbashree in Kerala." *Innovation and Development* 3(1): 139–140.

Collins, Daryl, Jonathan Morduch, Stuart Rutherford, and Orlanda Ruthven. (2009) *Portfolios of the Poor: How the World's Poor Live on $2 a Day.* Princeton, NJ: Princeton University Press.

Datta, Rekha. (2000) "On their Own: Development Strategies of the Self-Employment Women's Association (SEWA) in India." *Development* 43(4): 51–55.

Develtere, Patrick. (1993) "Cooperative Movements in the Developing Countries. Old and New Orientations." *Annals of Public and Cooperative Economics* 64(2): 179–208.

Du Bois, W.E.B. (1907) *Economic Co-operation among Negro Americans.* Atlanta, Georgia: Atlanta University Press.

———. (2007 / 1903) *The Souls of Black Folk.* Minneapolis, MN: Filiquarian Publishing.

Fairbairn, Brett. (1994) *The Meaning of Rochdale: The Rochdale Pioneers and the Co-operative Principles.* Occasional Paper, pp. 1–62.

Fatton, Robert. (2007) *The Roots of Haitian Despotism.* Boulder, CO: Lynne Rienner.

———. (2002) *Haiti's Predatory Republic: The Unending Transition to Democracy.* Boulder, CO: Lynne Rienner.

Geertz, Clifford. (1962) "The Rotating Credit Association: A Middle Rung in Development." *Economic Development and Cultural Change* 10(3): 241–263.

Gordon Nembhard, Jessica. (2014) *Collective Courage: A History of African American Cooperative Economic Thought and Practice.* University Park: Pennsylvania University Press.

Grant, William, and Hugh Allen. (2002) "CARE's Mata Masu Dubara (Women on the Move) Program in Niger: Successful Financial Intermediation in the Rural Sahel." *Journal of Microfinance* 4(2): 189–216.

Gray, Obika. (2004) *Demeaned but Empowered: The Social Power of the Urban Poor in Jamaica.* Kingston, Jamaica: University of West Indies Press.

Guinnane, Timothy. (2001) "Cooperatives as Information Machines: German Rural Credit Cooperatives, 1883–1914." *Journal of Economic History* 61(2): 366–389.

Handa, Sudhanshu, and Claremont Kirton. (1999) "The Economies of Rotating Savings and Credit Associations: Evidence from the Jamaican 'Partner.'" *Journal of Development Economics* 60(1): 173–194.

Harrison, Faye V. (1988) "Women in Jamaica's Informal Economy: Insights from a Kingston Slum." *New West Indian Guide* 3(4): 103–128.

Healy, S. (2009) "Economies, Alternative." In R. Kitchin and N. Thrift (eds.), *International Encyclopedia of Human Geography*, pp. 338–344. Oxford, UK: Elsevier.

Heinl, Robert Debs, and Nancy Gordon Heinl. (2005) *Written in Blood: The Story of the Haitian People 1492–1995.* Lanham, MD: University Press of America.

Hossein, Caroline Shenaz. (2020) "Rotating Savings and Credit Associations (ROSCAs): Mutual Aid Financing." In J.K. Gibson-Graham and Kelly Dombroski (eds.), Section 5 in *The Handbook of Diverse Economies*, pp. 354–361. Cheltenham, UK: Edward Elgar Press.

———, editor. (2018a) *The Black Social Economy in the Americas: Exploring Diverse Community-Based Markets.* New York, NY: Palgrave Macmillan.

———. (2018b) "Banking While Black: The Business of Exclusion." *Conversation.com* May 7, https://theconversation.com/banking-while-black-the-business-of-exclusion-94892

———. (2017) "Fringe Banking in Canada: A Preliminary Study of the 'Banker Ladies' and Economic Collectives in Toronto's Inner Suburbs." *Canadian Journal of Non-profit and Social Economy Research* 8(1): 29–43. Open access: https://anserj.ca/anser/index.php/cjnser/article/view/234

———. (2016) *The Politics of Microfinance: A Comparative Study of Jamaica, Guyana and Haiti.* Toronto, ON: University of Toronto.

———. (2015) "Government-owned Micro-Bank and Financial Exclusion: A Case Study of Small Business People in East Port of Spain, Trinidad and Tobago." *Canadian Journal for Latin American and Caribbean Studies* 40(3): 393–409.

———. (2014a) "The Exclusion of Afro-Guyanese in Micro-Banking." *The European Review of Latin America and Caribbean Studies* 96(1): 75–98.

———. (2014b) "Haiti's *Caisses Populaires*: Home-grown Solutions to Bring Economic Democracy." *International Journal of Social Economics* 41(1): 42–59.

———. (2013) "The Black Social Economy: Perseverance of Banker Ladies in the Slums." *Annals of Public and Cooperative Economics* 84(4): 423–442.

James, Carl, David Este, Wanda Thomas Bernard, Akua Benjamin, Bethan Lloyd, and Tana Turner. (2010) *Race and Well-being: The Lives, Hopes and Activism of African Canadians*. Halifax, Canada: Fernwood Publishing.

James, C.L.R. (1989) *The Black Jacobins: Toussaint L'Ouverture and the San Domingo Revolution*. 2nd ed. New York, NY: Vintage.

Maguire, Robert. (1997) "From Outsiders to Insiders: Grassroots Leadership and Political Change." In R. Maguire (ed.), *Haiti Renewed: Political and Economic Prospects*, pp. 154–166. Washington, DC: Brookings Institution.

Mayoux, Linda. (1999) "Questioning Virtuous Spirals: Microfinance & Women's Empowerment in Africa." *Journal of International Development* 11(1): 957–984.

Mendell, Marguerite. (2009) "The Social Economy of Quebec: Lessons and Challenges." In D. Reed and J.J. McMurtry (eds.), *Co-operatives in a Global Economy: The Challenges of Co-Operation across Borders*, pp. 226–242. Newcastle-upon-Tyne, UK: Cambridge Scholars Publishing.

Mensah, Joseph. (2010) *Black Canadians: History, Experience, Social Conditions*. 2nd ed. Halifax, Canada: Fernwood Publishing.

Mintz, Sidney. (2010) *Three Ancient Colonies: Caribbean Themes and Variations*. Cambridge, MA: Harvard University Press.

———. (1955) "The Jamaican Internal Marketing Pattern: Some Notes and Hypotheses." *Social and Economic Studies* 4(1): 95–103.

Montasse, Emmanuel. (1983) *La Gestion Strategique dans le Cadre du Dévelоppement d'Haiti au Moyen de la Coopérative, Caisse d'Epargne et de Credit*. Port-au-Prince, Haiti: IAGHEI, UEH.

Nelson, Nici. (1996) "The Kiambu Group: A Successful Women's ROSCA in Mathare Valley, Nairobi (1971 to 1990). In S. Ardener and B. Sandra (eds.), *Money-Go-Rounds: The Importance of Rotating Savings and Credit Associations for Women*, pp. 49–71. Oxford, UK: Berg.

Niger-Thomas, Margaret. (1996) "Women's Access to and the Control of Credit in Cameroon: The Mamfe Case." In S. Ardener and B. Sandra (eds.), *Money-Go-Rounds: The Importance of Rotating Savings and Credit Associations for Women*, pp. 95–111. Oxford, UK: Berg.

N'Zengou-Tayo, Marie-José. (1998) "*Fanm Se Poto Mitan*: Haitian Woman, The Pillar of Society." *Feminist Review: Rethinking Caribbean Difference* 59(1): 118–142.

Paradkar, Shree. *Toronto Star*. (Feb 6th 2020). "This Egyptian-Canadian woman went to withdraw her own money at RBC. What Happened next is the subject of a lawsuit against the bank and Peel Police." Open access: https://www.thestar.com/news/gta/2020/02/06/this-egyptian-canadian-woman-went-to-withdraw-her-own-money-at-rbc-what-happened-next-is-the-subject-of-a-lawsuit-against-the-bank-and-peel-police.html

*Poto Mitan: Haitian Women, Pillars of the Global Economy*. (2008) Film; 60 minutes. Prod. Tet Ansanm. www.potomitan.net/

Rutherford, Stuart. (2000) *The Poor and their Money*. New Delhi, India: DFID / Oxford University Press.

Sethi, Raj Mohini. (1996) "Women's Informal Cooperative Banks in Contemporary Indian Society." In S. Ardener and B. Sandra (eds.), *Money-Go-Rounds: The Importance of Rotating Savings and Credit Associations for Women*, pp. 163–179. Oxford, UK: Berg.

Sives, Amanda. (2010) *Elections, Violence and the Democratic Process in Jamaica: 1994–2007*. Kingston, Jamaica: Ian Randle.

St. Pierre, Maurice. (1999) *Anatomy of Resistance: Anticolonialism in Guyana 1823–1966*. London, UK: MacMillan Education.

Sterritt, Angela. (9 January 2020). CBC. Indigenous grandfather and 12-year old grand-daughter handcuffed in front of a Vancouver bank after trying to open a bank account. Open access: https://www.cbc.ca/news/canada/british-columbia/indigenous-girl-grandfather-handcuffed-bank-1.5419519

Tafari-Ama, Imani. (2019) *Blood Bullets and Bodies: Sexual Politics below Jamaica's Poverty Line*. Lancashire, UK: Beaten Track Publishing.

Thiagarajan, Kamala. (4 January 2019) "Millions of Women in India Join Hands to Form A 385-Mile Wall of Protest." *NPR Radio*.

*Toronto Star*. (10 December 2014) "Banking While Black." Retrieved 7 August 2015: www.thestar.com/news/gta/2014/12/10/banking_while_black_toronto_man_accuses_scotiabank_of_ racial_profiling.html

Washington, Booker T. (2013/1901) *Up from Slavery: An Autobiography*. Delhi, India: Ratna Sagar Press Ltd.

Witter, Michael. (1989) "Higglering/Sidewalk Vending Informal Commercial Trading in Jamaican Economy." Occasional paper series 4(6). Department of Economics, University of West Indies / Mona.

Wong, David. (1996) "A Theory of Petty Trading: The Jamaican Higgler." *Economic Journal* 106(3): 507–518.

## Interviews and focus groups

"Betty," Interview (Senior community activist and participant in an informal cooperative), Toronto, 31 July 2015.

Bon Repos, Focus groups, Haiti, 9 October 2010.

"Fardowsa," Focus group, Toronto, 20 March 2015.

"Faye," Focus group, Toronto, 20 March 2015.
Jane and Finch, Focus group, (20 people), Toronto, 20 March 2015.
"Mabinty," Focus group, Toronto, 20 March 2015.
Member of an informal cooperative bank, Interview, Kingston, July 2009.
"Miss Paddy," Interview (Member of an informal cooperative bank), Kingston, Jamaica, 6 May 2009.
"Natla," Focus group, Toronto, 26 March 2015.
"Rickie," Interview (Bar owner and member of an informal cooperative bank), Kingston, Jamaica, 9 July 2009.
Scarborough, Focus group (26 people), Toronto, 26 March 2015.
Senior Banker, Interview, Port-au-Prince, Haiti, 2 October 2010.
Three informal bankers, Interview Kingston, Jamaica, March to July 2009.

# 7 Rondas Campesinas and Defense Fronts

## The environmental justice movement confronting extractivist policies in Peru

*Raquel Neyra*

### Unsustainable exploitation of Nature

According to defenders of the varieties of capitalism approach – whether they articulate neoliberalism, progressivism, or the national economic development approach – growth will eradicate poverty through national economic enhancement. Within the global South we are pushed incessantly towards new sources of raw materials, minerals, hydrocarbon, water, marine biomass land for intensive agriculture, and livestock (Gudynas, 2013a, 2013b). Unsustainable extraction limits are extended over the continents, to past colonies (Moore, 2013; Conde, Walter, 2015).

Socio-economic metabolism is the ecological concept that traces human controlled material and energy flows that occur between Nature and society. It constitutes the self-producing biophysical basis of society. Mounting changes in such social metabolism (Martínez Alier et al., 2016) push us to greenhouse effects, drought, and abnormal rains, amidst the lures and hazards of profit-making. Specifically, we focus on the extractivist industry as it pushes ecological limits in exploiting natural resources in the drive to sell on the world market (Moore, 2013; Conde, Walter, 2015). We are increasingly confronted with unsustainable consumption of material and energy, which will not be replaced or be recycled. This chapter studies a pushing back by peasants and Indigenous peoples (Figure 7.1).

With the advances of technology, shovel extractors and rigs mine more in less time. More and more land is removed, more and more rivers are diverted. The food sovereignty of the South of the Planet is threatened. Wastes from extractivist industries still remain in the exploited locations, causing deaths and illness. Here in Peru, thousands of people are contaminated like the *niños del plomo* (lead kids) in the

*Figure 7.1* Meeting of Rondas Campesinas commemorating the victory over the mining project in the town of Unión Las Minas, district of Tabaconas, San Ignacio province in the foothills of the Andes in Peru

Source: Raquel Neyra.

seaside city of Callao, which is the port of Lima, the same is seen in La Oroya or Cerro de Pasco. Those children embody lead and mineral levels over the maximum level permissible (Arauzo, 2009). The countless oil spills in Amazonia pollute ever more water and food sources of Indigenous people – nearly 20 spills between 2016 and 2017 in the area west of Brazil that this chapter studies, the highlands of Northeast of Peru (Amnesty, 2017). This amounts to the gradual erasing of a people.

## Two types of social community movement organizations going up against extractivist policies in northeast Peru

In the 1990s, the neoliberal policies of the Peruvian dictator President Alberto Fujimori (1910–2000) focused on extractivism as the foundation of all the future governments. Peasants and Indigenous people were faced with the new threat of the presence of extractive companies in mining, oil, timber, and monoculture agriculture companies in the lands they worked and inhabited. Little by little, these people resist and organize themselves. They manifest warranted assertions of justice amidst the first steps of the extractive activity – land grabbing, then water grabbing, and in the end environment pollution. Movements for environmental justice are not new. We can go back in Peru to 1934 to such movements in San Mateo de Huanchor. But more recently, they have been seen to multiply as they respond to accelerating extractivism. In response, local SCMOs emerge in defense of the territory, such as the *Frentes de Defensa* (defense fronts) and *Rondas Campesinas* (peasant patrol rounds).

In the 24 departments of the country of Peru we find Indigenous populations, in peasant and native communities. The recognized and titled peasant communities (5,137) occupy more than 24 million hectares; of these, another 2,130 communities are reported "missing yet recognized by owner" (IBC, CEPES, 2016). To these, add 1,359 native communities amounting to titled extension of more than 12 million hectares; of these, 800 communities are reported missing yet recognized by owner. This area, titled and recognized by owner, occupies more than 50% of the total area of the country (IBC, 2016). This territory is inhabited by 55 Indigenous nations with 47 languages (Ministry of Education, cited by the Office of the Ombudsman) representing four million people (INEI, 2017).

Community movement organizations emerge here, together with others from civil society. They unite to develop protest actions and make visible the reality of conflicts caused by exploitation beyond the limits of sustainability. Behind a common goal and the recognition of a collective identity (Diani, 1992), these Indigenous and peasant SCMOs constitute a resistance movement. Their collective action is of a kind that is not going to be integrated in the existing institutionalized context. Rather, they are going to constitute a novel connecting-ness in response to the extractivist threat. This is a response constituted in preferences to change the social structure or distributive ecological conflicts: a response to make visible the incompatibilities within the system, ones that surpass the limits of the system (Diani, 1992 citing McCarthy and Zald, 1977).

What we will see in this movement response are the assemblages of interwoven *rhizomatic practices* that thread the case studies of this anthology: an emergent meshwork of interconnected performances expressing mutual commitment and accountability. This reinforces the development of a notion of reciprocal solidarity held together by shared values (Melucci, 1989, cited by Diani, 1992) in the struggle against extractivist exploiters. Social movement organizations constitute alliances with other local movement organizations to be able to advance their warranted assertions and, in the case of Peru, defend their lives.

This chapter presents two types of organizations that resist against extractive activities, as a concurrence of *rhizomatic emergence* – making explicit the incompatibilities in the social structure and the distribution conflicts they engender. Sometimes, a Defense Front calls itself *Platform* as in Celendín, Cajamarca, e.g., Plataforma Interinstitucional Celendina that resisted against the Conga Mining project. The author has been a member of this Platform, as well as the NGO research project Envjustice / EJ Atlas.

The *Rondas Campesinas* are to a large extent self-organized alternatives to practices associated with capitalist justice forms. However, with the advance of extractivism, they resist extractive activities, defending their way of life based with *buen vivir* forms of life – forms that colonialism, coloniality, and republican epochs have partially destroyed. In this sense, Rondas Campesinas apply mutualistic methods as *minga* (work today for me, tomorrow for you) or *ayni* (solidarity, reciprocity), two components of *buen vivir*. These two words are from the *quechua* language. Other Indigenous people may use words in their own language.

On the other side, the Frentes (Defense Fronts) are horizontal models of self-managed representation. They act as an alternative of management facing governmental policies. They observe the state's shortcomings or everything that threatens the people's integrity as extractive industries and claim against it.

Both are informal cooperative arrangements, completely horizontal, that challenge capitalistic hierarchical structures (here capitalist structures of coloniality and modernity), implying a radical critique of the capitalistic system of organization and distribution.

In Peru we do not share European or American mentalités and practices; we are always in deep reality constructions of coloniality. The anti-extractivist SCMOs of the Andes people are not exactly "prefigurative politics." These forms of social self-management were practices before colonialism. They survive the colonial and republican epochs. They are "recovered."

### The Defense Fronts

*Frentes de Defensa* are a meshwork of "fronts" knit together by SCMOs of a locality, province, or region. Generally, they are interlaced in a connectingness to defend common interests, to sustain common goods (Ramos, 2012; Neyra, 2013a, 2013b). Some of them specify the sphere of action, such as "environmental defense front," but mostly called themselves "of the interests." The Fronts serve as a functional alternative where the State is absent or where it fails to provide regulative or resolutive actions at local or regional level. The Fronts confront the problems that occur due to the presence of extractive industries. The Fronts serve to watch out for the interests of The People in healthcare, economic labor, and environmental and neighborhood affairs.

Often, the State acts and presents itself as weak, pretending that it has not enough resources to supply/cover the population's needs. But in fact, it is representing the companies. In this way, the State can press more and open the way for corruption and the control of the State by the extractivist companies. The latter's public relations purpose is to offer to cover population needs, seeking to attract The People to their side.

One of the main particularities of the Fronts is their connecting-ness. They establish alliances with other guilds and associations of civil society as well as social movement organizations to maximize their claims. The Fronts knit themselves in popular meshwork together with trade unions, student federations, religious groups, as well as neighborhood associations of producers, farmers, merchants, and peasants. What draws them together is a perceived threat from privileged economic forces, e.g., students against pollution, farmers and peasants against loss of land, and merchants who cannot sell polluted products.

- The earliest significant Fronts in Peru were constituted against the imposition of *scholar fees* in Ayacucho region (Ramos, 2012). Householders and students joined together to claim their social right to higher education. Incrementally, the action sphere of the Fronts was expanded to include economic, social, and political rights.
- Later, in the 1990s, some of the names of the Fronts include the adjective "patriotic," e.g., the Loreto Patriotic Front. In that case, the Loreto teachers protested the government intention to revise the Río de Janeiro Protocol of 1942 that sealed the end of war and demarcated the frontiers between Peru and Ecuador (Ramos, 2012).
- In the Amazon case, the Defense Fronts demonstrated against the intervention of hydrocarbon, oil and mining companies, as well as against the mega-dams and against inter- oceanic path construction.
- In the Andes Zone, the Environment Defense Fronts were created[1] to resist mining projects and their attempt at expansion.

The Defense Fronts' decisions are taken in open assemblies where all community organizations and their representatives participate. They apply horizontally principles of participatory democracy. They identify a common interest that they consider injured. Decisions are made on the basis of majority rule, which must be respected. The stratagem used by the Fronts starts with the editing of claims that land rights have been violated and handed over to the extractivists. These are followed by peaceful demonstrations. If their demands are not heard or resolved then the Defense Fronts promise to radicalize their actions with strikes, mass demonstrations, contestation of public space, destruction of symbols, road blocks, and occupation of the companies' buildings/corporate offices. (Those of us working at the Envjustice Project/EJAtlas have observed some of these cases.) Governments often answer with repression and violence. This was the case of Tía María Project with five people shot by the police. There were two people shot in Pichanaki; five people killed by the police by Las Bambas mining project; and five people assassinated during the protests against the Conga project (Jesamine, 2016).

But, on the other side, where the Fronts have been well organized and strong, they have been successful, e.g.,

- the Saramurillo conflict (2016, 2017) where the Front blocked transport on Amazonian rivers;
- the Pantanos de Villa conflict (2003) that united all of civil groups of the town;
- against the Conga mining project (2013) that mobilized thousands of people and associations around many provinces.

The Fronts monitor the fulfillment of the State's obligations in vital services like health, education, etc. In doing so, the Fronts bring their members and meshwork allies together to stand as a watchful force with strong capacity. They mobilize to defend and preserve their habitat, fighting for more environmental justice against the interests of extractivist companies and against the State's complicity with the extractivists. They represent what are now labeled *ecologial distribution conflicts* (Martínez-Alier, 1999), i.e., conflicts over the appropriation of Nature, over the distribution of wastes and polluting zones (see EJAtlas).

### The Rondas Campesinas

The *Rondas Campesinas* are peasant community organizations that began to emerge in 1976 in the Cajamarca region of Peru, in the middle of the military dictatorship of President Morales Bermudez. The country was in a major recession – wracked by economic crises of debt, spiraling inflation, and money that was devaluated. Unemployment increased – exacerbating insecurity and affecting the entire country.

The emergent Rondas began as a response to their witnessing limitations to land reforms brought about in the late 1960s, and a sense that social order was not being provided by the State (Fumerton, 2001). It was also a time of a lot of robbery as Andean peasants were victims of livestock theft. The Rondas Campesinas organized to fight against such robbery[2] – to mutually administer justice by themselves. A felt lack of institutional presence obliged the Rondas to organize and to defend themselves against any intrusion in their territory, mostly against cattle thieves.

Confronting an epidemic of cattle rustling, highland farmers organized nightly civil patrols. The Rondas were the *de facto* authorities judging thieves, administering all the justice in their communities (Starn, 1999). The *Rondas Campesinas* came to be officially recognized by the State in the article 149 of the Peruvian Constitution. But since their creation, many governments tried to withdraw its powers from the Rondas.

Specifically, the Rondas were founded by the Andean peasants in Cuyumalca – a little town in the district and province of Chota, in the Cajamarca region – on December 29th 1976 as a completely autonomous SCMO (Korbaek, 2013). Step by step, the action sphere of the Rondas expanded to all concerning members of the community. Anyone can be a member of the Rondas, including teachers, merchants, and others; they are called *rondero/rondera*.

The word *Rondas* comes from the principal activity that the *Ronderos* practice: *rondar,* which means the "rounds" taken to patrol and to keep watch through the fields and towns. The Ronderos monitor suspicious behavior or uncommon persons in the communal territory. Most of the time, they walk in the night (see again Starn, 1999). Their meetings take place at strategic points and in turns – watching all the communal territory. They are authorized to apprehend any person that moves through the territory, to question her, and to lead her to the *Casa Rondera* if they are not convinced by the justification proffered by the stranger traversing the communal territory. The *Rondas house* is a small building, most of the time just a room, where the Ronderos spend the night and come together.

The community members come together and elect by quorum the *Ronderos* in an open vote. *Ronderos* do not work as wage earners. The decisions are taken in assemblies where all the community participates. The most significant characteristic is that they administer their own justice. Justice is applied immediately and is not appealable. This is known as *rondera justice.* The punishment consists in communal work. The only objective of the punishment is to give harmony back to the community. People are sometimes punished with a whip.

Over time the action sphere of the *Rondas* expanded to every offence, like land, borderland or neighborhood troubles; inappropriate behavior of any community member; family problems; and rapes, physical violence, and assassinations (Sánchez Ruiz, 2014). This courageous territory surveillance stopped the activities of the terror groups "Sendero Luminoso" (Shining Path) and "MRTA"[3] in Cajamarca. These groups began their acts of intimidation in year 1980, wanting to coopt people into their terrorist guerrilla movements (Gitliz, Rojas, 1985). The patrols also counter the counterinsurgency efforts by the central government in Lima (Fumerton, 2001).

The communal decision-making and the application of its own justice can be considered as a part of *desborde popular*, i.e., popular overflow (Matos Mar, 1986). This means that when people overturn the inefficiency or nonexistence of the State functioning, people assume the role that the State has been unable to exercise. People here – long too tired waiting, and long too tired to claim – finally arise to take justice, to empower themselves. Like the Fronts, the Rondas arise instituting alternatives to an instituted deficient State. They too react to the deteriorated and substandard life condition.

It is worth noting here that, for some years now, there have been "Urban Rondas" in Cajamarca. Their role resembles that of a neighborhood council; they can arrest people who commit an offence and deliver people to the police.[4]

The *Rondas campesinas* are organized in Committees that integrate Federations at local, province, or regional level. Its representation at national level is the CUNARC, *Central Única Nacional de Rondas Campesinas*. It is estimated that actually there are nearly 250,000 *Ronderos*, of which 100,000 are in Cajamarca, regrouped in 8,000 comités (Korsbaek, 2013). Unfortunately, the political interference of parties creates division in Cajamarca; there exist now the *Rondas Campesinas Federation* and the *Rondas Central Committee*.

Female-based *Rondas* were created in 1988 in Cajamarca. Women organize to fight against machismo, and their ranks steadily increase. Unfortunately, the *Ronderas* are chided by others because of their "absence in the household." Despite their key role, the *Ronderas* are not encouraged to assume a regional or national leadership (Neyra, 2015).

Given the practice by Rondas of apprehending strangers in its territory, the Rondas will act against the presence of mining companies and other extractives activities. This was especially true after 1992 Peruvian government concessions to profit-driven multinational corporations from the North and China. Extractive activities enabled by these Peruvian government concessions feed cycles of unsustainable consumption in globalized capitalism. Such extractive activities are understood by the Rondas as intrusion in communal territory. When they defend their environment, they defend not just their ways of life but life itself. Open pits, mining camps, mining roads, field destruction, pollution, and wastes are considered violations of their territory.

A lot of resistance movements against extractivist projects would not have been possible without the active intervention of the Rondas. *Ronderos often paid for this intervention with their lives.* Ronderos continue to lead the resistance in Majaz, San Ignacio, Tabaconas, Conga, Chadín II, and many other extractive projects or mega dam projects (EJatlas, 2019). The Rondas still have a major role in the resistance against the Conga Mining Project or the Lagunas Norte Barrick Gold´s project (La Libertad region), against mining project from Canadian (Candente Copper) or Chinese (Zijin) mining companies, and many more. With their resistance, they question the globalized capitalism model and bring to light the ecological distribution conflicts and property conflicts over goods and territory (Martínez Alier et al., 2016).

Unlike a Defense Front, the Rondas recognize themselves as a collective identity: in the sense of a shared culture, a shared historicity, and a shared definition of themselves. This shared collective identity warrants their power to act over all the community. However, just like the Fronts, the Rondas represent organized civil society in assuming decisions in an autonomous and self-managed way. Each acts as they consider appropriate for the defense of their wellness. Rondas and Fronts institute themselves; they recover and build political space.

Both reject the destructive extractivism in their own country and in the world. Both opposed the neoliberal austerity measures under the regime of President Fujimori that began in the 1990s. They set itself up as an instituting force, but not to replace an absent State. Quite the contrary, to replace a too much present and instituted State with the police presence and law enforcement in favor of extractives companies. At the same time, they disrupt capitalist production and distribution chains (Acosta, Brand, 2017).

The Frentes use a confrontational repertoire of action, while the Rondas more approximate SCMOs as "practice based movement organizations." The latter organizations have at their core the sentiment and purpose of replacing unsustainable practice and forging alternative, productive, and sustainable flows and institutions. The Rondas have non-self-consciously engaged in building alternative practices of replacing unsustainable habits. The Rondas are not cognitively driven; theirs is a felt movement. They know now what buen vivir means; they resist against their destruction. They resist with limited means, as they are poor or very poor. They forge alternative, productive, and sustainable flows in few cases, mostly with the help of other organizations, such as NGOs or the local government. Their experience is one that Touraine (1977) refers to as *rapports* of felt immersion: intersubjectively acknowledging one another as co-authors of practices of de-coupling and re-institutionalizing.

On the other side, the Fronts have the idea of replacing unsustainable practice like the Rondas, but not everyone in the Fronts talks about building alternatives. That has not been their primary objective. They see themselves more as "watchmen" (Starn, 1999). To speak about prefiguratively forging alternative, productive and sustainable flows and institutions for both may be too early. As SCMOs, the two are mutualistic experiences of welfare from below addressing real needs; and are horizontal models – as informal cooperative – arrangements employing a radical critique of capitalist forms and practices.

## Indigenous peoples of the Andes and *Buen Vivir*

Often, the Indigenous Peoples of the Andes do not understand the imbroglio of capital relations they are being entangled into. For example, there is the case of the peasants in Las Bambas project. First, they approve the mining project but later, they suffer under pollution and state violence. It is then, that they have understood what is really going on – how their nation, their culture is threatened; how their existence itself is threatened.

In all environmental justice struggles in Peru, the social movement organizations that defend the environment are led by Indigenous people or by the affected inhabitants. Indigenous people led the resistance in 90% of the resistance struggles.[5] This fact could be better appreciated by the rest of the Peruvian population as the Indigenous people are still underestimated.

The struggles against extractive activities amount to a defense of environment. It leads them to search in their heritage for a reason to exist, a way of life to defend – far away from capitalist consumption societies models. And this means less consuming of unnecessary things. The Spanish invaders tried to annihilate their cultures – from North to South America. But they didn't destroy all of the material practices. There remained some kernel of communal life around "buen vivir."

*Buen vivir* – living well – is the Spanish term for the Indigenous worldview of *sumac kawsay*: i.e., ecologically balanced community life, collaborative consumption, and a sharing economy. The worldview (*Weltanschauung*) is a culturally sensitive and community-centric way of doing things – not about the individual *per se* but about the individual in both a social solidarity context and a respective unique environmental situation. Humans are understood not as owners, but as reciprocating stewards of the Earth and its resources (Gudynas, 2013a, 2013b).

Living well or *sumak kausay* in *kichwa* idiom is a holistic view of life that proposes to live in harmony and balance with the cycles of Mother Earth, of the cosmos, with all forms of existence. As a world view, it respects the diversity existing within Nature and we as part of diversity, that is to say, a total respect for the different peoples and that respects their autonomy (Huanacuni, 2010). It does not consider Nature as natural capital, but as a being without which life does not exist. For this, we must maintain the balance, that is, preserve the integrity of the natural processes that guarantee the flow of energy and materials in the biosphere and preserve the biodiversity of the planet. That said, every form of aggression – such as extractivism – that does not reproduce life is contrary to living well. Living well involves practices of production and exchange called *minga*, a system of reciprocity. It is a logic of exchange that leads to a solidarity economy. Indeed, such solidarity economy is only possible with *buen vivir* community life.

The community is anchored in a relationship between being and territory, between being and Nature. These attitudes of life do not then promote consumption fueled by extractivism nor of exchange in terms of capitalist markets. In its holistic and integrative worldview, there is no concept of development, because there is no linear Eurocentric concept of history. There is not a before or an after or a state to be achieved. A concept of poverty associated with the lack of material goods or wealth associated with the abundance of these goods is unknown (Acosta, 2017). Their codes of conduct are based on respect for knowledge, one that involves both ethical and spiritual behaviors founded in the society-Nature relationship.

*Buen vivir* is not a discursive tool coopted in a functional sense by capitalism or by the State. *Buen vivir* is appropriated in a post-development market-driven sense of "natural capital" – in a continued project of the interchangeability of capital. But monetary value cannot be given to environmental goods like the water provision of rivers or the removal and storage (e.g., sequestration) of carbon from the atmosphere of forests and soil. Nor is *buen vivir* understandable in a state socialist governmental sense of the likes of Correa in Ecuador and Morales in Bolivia – as a policy of harmonious living together in sustainable development (Walsh, 2010; Merino, 2016).

People who defend their habitat and way of life (habitus in the Bourdieusian sense) participate in the construction of the *buen vivir*, the respect for the Nature to which we belong, as a Communal being where we are, and not a being outside of us. They defend, not an archaic or idyllic way of life, but a way of life where Nature has a primary role, where communion with the being's relations is fundamental.

All cultures that existed before the Spanish colonial epoch move in symbiosis with Nature. They move around Nature. All cultures have a *cosmovision*, understanding how every part of its cultural whole is related to Nature. We are she, she is us. A non-rational relation, a non-abstract relation, is created. It is an affective relation, what Raymond Williams (1954) once referred to as a "structure of feelings." It is a perceived relation, where environment is absorbed and where there is an interchange of forces. Arturo Escobar (2014) develops the expression *sentipensar* "feel-think" in trying to transmit this mode of perception.

There exist in Peru good initiatives to rescue and spread the concept of *buen vivir*. For example, with the help of some NGOs, the Awajun community[6] has established a document about *buen vivir* and made a video. They have understood that their *cosmovision* and way of life have helped to conserve and reproduce the Amazon forest for centuries. Like the Rondas Campesinas, the Awejun share a collective identity to which they refer to, engaging themselves in ecological struggles. The *buen vivir* is part of this collective identity, and thus became the reason to resist. Andean or Amazonian social movements as the Rondas Campesinas or some Defense Fronts construct a link between environmental justice and the world view of *buen vivir*.

Indigenous people of the Andes sustain biodiversity; they take advantage from the vegetation and conserve them too. And they adapt their housing to the natural environment. They move and act around the "local" – i.e., in the surrounding territory – and live in harmony with it. Their production forms have been ecological and renewable. We can appreciate how they built the "andenes" (agricultural terraces) that preserve water and soil, produce different foods, and avoid erosions and droughts (Diestel, 2018). They resist dogmatic preaching of the need to ever-increase – for growth as the vital core development factor. Their resistance does intend physical conflict with the interest of extractive companies and the State, who nonetheless react with notable violence.[7] It doesn't mean that we should go back to a supposed idyllic past, but, like Aníbal Quijano (2006) says, we should reinvent a new world between our past and today's culture.

## Conclusion: reciprocal solidarity

The Rondas and the Defense Fronts seek to reach *well living / buen vivir* and to respect the environment. They resist the ways of capitalist production, extractivism, exploitation of human beings, and territory appropriation. As part of both the social solidarity economy movement and the environmental movement, they defend common goods, and show that it is vital to reduce voracious economic growth. They construct a link between traditional knowledges and the preservation of the Nature (Villamayor-Tomas, 2018). When they resist, they seek a sustainable way of life; they act for the sustainability of the Earth. Issues of biodiversity and climate change involve *reciproqueteurs* rather than *entrepreneurs* (Weiner, 2018). This is a willingness to sacrifice some material self-interest in order to sustain that which redounds to all participants in a meshwork and involves sanctions, discipline, and punishment of opportunistic meshwork participants.

This is a beginning. The quest for *buen vivir* needs to grow more and more. It should articulate itself experimentally and more concretely. It should interrelate with other environmental justice movements around the world in the struggle for a better world, for life itself. This involves a strong necessity to create meshworks, and that will not be easily given the physical distance beyond digital space as well as the differences in languages and mentalities. This is a challenge for creativity, for a new imaginary. Movement organizations like Rondas and Defense Fronts have understood that the only way to survive is to cooperate.

In Europe and in a part of the North, the destruction and annihilation of the cultures before capitalism was almost total, except few practices that subsist among native American Indians, Australian aborigines, or Lapland natives. Furthermore, those expressions that subsist aim, by their presence, for a resurgence organized around environmental justice objectives: Sioux people demonstrate against a pipeline through their territory. Australian aborigines and Lapland people protest the reduction of their territory that is threatened by the "civilization" (e.g., Norwegian trains overwhelm reindeer herds).

Social solidarity movements and environmental justice movements defend common goods. They understand that the only way to survive is to cooperate. They are part of the *buen vivir*. They construct a link between the material practices of traditional knowledge and the preservation of Nature (Villamayor-Thomas, 2018). Where they resist, they seek both a sustainable way of life and a sustainable Earth. There is a need to construct meshworks on multiscalar levels – not so easily done, given the physical distance beyond digital space as well as the differences in languages and mentalities. This is a calling that radiates from within the forming of assemblages of meshwork connectivity, a challenge to a 21st century institutionalizing imaginary (see Castoriadis, 1987).

## Notes

1 I am a member of Frente de Defensa Ambiental de Cajamarca.
2 Cattle thieves.
3 MRTA: Túpac Amaru Revolutionary Movement.
4 Municipal Ordinance no390 CMPC, Municipality of Cajamarca, June 27, 2012.
5 See EJAtlas.org.
6 See different initiatives on Tajimat Pujut (good living), on YouTube or in writing.
7 According to Global Witness, in 2014 Peru was the fourth country with the most deaths of environmental defenders, only in 2015 were 12 people killed and in 2016 two people. But the figure may be higher because not all cases are known.

## Bibliography


Acosta, Alberto, and Brand, Ulrich, 2017. *Salidas del laberinto capitalista, Decrecimiento y Postextractivismo*, Rosa Luxemburg Fundation, Ecuador.

Amnesty International, 2017. *Estado tóxico, Report AMR* 46/7048/2017.

Arauzo, Godofredo, 2009. Contamination of La Oroya – Peru, *LEAD Action News*, 9(2), April 2009 ISSN 1324–6011.

Castoriadis, Cornelius, 1987. *The Imaginary Institution of Society*. Trans. Kathleen Blamey. London: Polity Press.

Conde, María, and Walter, Mariane, 2015. Frontières de la marchandise, en « Décroissance, vocabulaire pour une nouvelle ère » *compiled por G. Kallis, F. Demaria, G. D'Alisa, Ed. Le passager clandestin*, 2015, Neuvy en Champagne, France.

Coxshall, Wendy, 2016. 'When They Come to Take Our Resources': Mining Conflicts in Peru and their Complexity, *Social Analysis*, 54(1): 35–51.

Dedeauwaerdere, Tom, 2005. From Bioprospecting to Reflexive Governance, *Ecological Economics*, 53(4): 473–491.

Diani, Mario, 1992. The Concept of Social Movement, *The Sociological Review*, 40(1): 1–25.

Diestel, Heiko, 2018. *Zukunftsfähige Wasser- und Landwirtschaft zwischen Pazifikküste und Andengipfeln in Peru*, Conference by LAI-Freie Universität, 15/02/2018, Berlín.

Ejatlas, 2019, Callao, Peru, and Lead Pollution (2014), Canariaco Norte – San Juan de Kanaris, Peru, (2014), Cerro de Pasco, Perú (2014), Conga, Peru (2017), Derrames del oleoducto de Petroperú en la Amazonia, Perú (2016), Explotación maderera en los Bosques de San Ignacio, Peru (2017), La Oroya, Perú (2016), Las Bambas mining, Perú (2016), Luksic's investment damages Ramsar site of Pantanos de Villa, Lima, Peru (2017), Megarepresas sobre el Marañon, Chadin 2 y Rio Grande 1 y 2, Perú (2016), Minera Las Palmeras en el Santuario Nacional de Tabaconas en Cajamarca, Peru (2016), Pichanaki and Pluspetrol, Peru (2016), Proyecto Lagunas Norte, Barrick Gold, Peru (2016), Rio Blanco Mine Majaz/Rio Blanco Copper S. A., Peru (2015), San Mateo de Huanchor, Perú (2014), Tia Maria, Perú (2014), https://ejatlas.org/country/peru

Escobar, Arturo, 2014. *Sentipensar con la tierra*, Medellín: Unaula.

Fumerton, Mario, 2001. Rondas Campesinas in the Peruvian Civil War; Peasant Self-Defense Organizations in Ayacucho, *Bulletin of Latin American Studies*, 20(4): 470–497.

Gitlitz, John y Rojas, Telmo, 1985. *Las rondas campesinas en Cajamarca-Perú, en: Apuntes N°16, Revista de Ciencias Sociales*. Universidad del Pacífico, Lima, pp. 115–141.

Global Witness, November 2014. *El Ambiente Mortal de Perú*, Report.

Gudynas, Eduardo, 2013a. *Extracciones, Extractivismos y Extrahecciones*: Observatorio del Desarrollo, CLAES, Montevideo, Uruguay.

———, 2013b. Interview; *Buen Vivir*: The Social Philosophy Inspiring Movements in South America, *The Guardian*, 4 February 2013.

Huanacuni, Fernando. 2010. *Buen Vivir / Vivir Bien: Filosofia, políticas, estrategias y experiencias regionales andinas. Lima: Coordinadora Andina de Organizaciones Indígenas (CAOI).*

IBC, CEPES, 2016. *Directorio Comunidades Campesinas del Perú*, SICCAM, Sistema de Información sobre Comunidades Campesinas del Perú, Lima.

IBC, 2016. *Directorio Comunidades Nativas del Perú*, SICNA, Sistema de Información sobre Comunidades Nativas de la Amazonía Peruana, Lima.

INEI, 2017. *Encuesta Demográfica y de Salud Familiar*. Nacional y Regional, Lima.

Jesamine, Cecelia, 2016. Community Opposition Forces Newmont to Abandon Conga Project in Peru, Mining.com. (18 April 2016).

Korsbaek, Leif, 2013. *Tipos de rondas campesinas en el Perú*, CUNARC Website http://cunarcperu.org

Jesamine, Cecelia, 2016. 'Community Opposition forces Newmont to abandon Conga Project in Peru," Mining.com. (18 April 2016)

Martinez Alier, J. Demaria, F., Temper, L., Walter, M., 2016, "Changing social metabolism and environmental conflicts in India and South America," *Journal of Political Ecology*, University of Arizona, EEUU,23(1)1,467–491,

Martinez-Alier, Joan, 2011, El ecologismo de los pobres: Conflictos ambientales y lenguaje de valoración, Barcelona; Icaria.

Martinez-Alier, Joan, 1999. Introducción a la economía ecológica, Barcelona: Editorial Rubes.

———, 2011. El ecologismo de los pobres: Conflictos ambientales y lenguaje de valoración, Barcelona; Icaria.

Martinez Alier, J., Demaria, F., Temper, L., and Walter, M., 2016. Changing Social Metabolism and Environmental Conflicts in India and South America, *Journal of Political Ecology*, University of Arizona, EEUU, 23(1): 467–491.

Matos Mar, José, 1986. *Desborde Popular y crisis del Estado*, Instituto de Estudios Peruanos, 3rd edition, Lima: Peru Problema 21.

Mccarthy, Joseph D., and Zald, Mayer, 1977. Resource Mobilization and Social Movements: A Partial Theory, *American Journal of Sociology*, 86(6): 1212–1241.

Melucci, Alberto, 1989. *Nomads of the Present: Social Movements and Individual Needs*. Philadelphia: Temple University Press.

Merino, Roger, 2016. An Alternative to 'Alternative Development'?: Buen Vivir and Human Development in Andean Countries, *Oxford Development Studies*, 44(3): 271–286.

Moore, Jason W., 2013. El auge de la ecología-mundo capitalista (I), Las fronteras mercantiles en el auge y decadencia de la apropiación máxima, *Laberinto* 38/2013, Málaga, Spain.

Neyra, Raquel, 2013a. Agricultura versus extractivismo La lucha por la supervivencia en el departamento de Cajamarca Perú, SEHA, Sociedad Española de Historia Agraria, Article for the XIV SEHA Congres: http://seha.info/congresos/artículos/CD.3.%20Neyra.pdf

———, 2013b. L'accaparement des terres au Pérou: les cas d'Olmos, de San Martin (Shawi) et de Conga, *Revue HISTOIRE(S) de l'Amérique latine*, Paris: Harmattan, Vol. 8, n° 7, 12 pp. May 2013, France.

———, 2015. Femmes dans la lutte: Rondas fémininas de Cajamarca et Ashaninkas contrel' Extractivisme (Pérou) Revista Caminando, n° 30, 2015, Québec, Canadá.

Quijano, Aníbal, 2006. El "movimiento indígena" y las cuestiones pendientes en América Latina, *Argumentos* [en línea] 2006, 19 (enero-abril): ISSN 0187–5795.

Ramos, José, 2012. Los Frentes de Defensa de los Intereses del Pueblo, su origen, significado y vigencia, in http://vanguardia-intelectual.blogspot.com.es

Sánchez Ruiz, Oscar, 2014. *Justicia Rondera*, Special edition para el XI Congreso de Rondas Campesinas y Urbanas de Cajamarca, 2014, Cajamarca, Peru.

Starn. Orin, 1999. *Nightwatch: The Politics of Protest in the Andes*. Durham: Duke University Press.

Touraine, Alain, 1977. *The Self Production of Society*. Trans. Derek Coltman. Chicago: University of Chicago Press.

Villamayor-Tomas, S., and García-López, G., 2018. Social Movements as Key Actors in *Governing the Commons*: Evidence from Community-based Resource Management Cases across the World, *Global Environmental Change*, 53: 114–126.

Walsh, Catherine, 2010. Development as *Buen Vivir*: Institutional Arrangements and (De) Colonial Entanglements, *Development*, 53(1): 15–21.

Weiner, Richard R., 2018. *Les Reciproqueteurs*: Post-Regulatory Corporatism, *Journal of Environmental Policy and Planning*, 20(6): 775–796.

Williams, Raymond, and Orrom, M. (1954). Film and the Dramatic Tradition, in R. Williams and M. Orrom, *Preface to Film*. London: Film Drama, pp. 1–5.

# Index

Note: **Bold** page numbers refer to tables; *italic* page numbers refer to figures and page numbers followed by "n" denote endnotes.

Printed in the United States
by Baker & Taylor Publisher Services